STUDENT HANDBOOK AND SOLU

HARRY NICKLA

Creighton University

GENETICS
A MOLECULAR PERSPECTIVE

William S. Klug
Michael R. Cummings

Prentice
Hall

Upper Saddle River, NJ 07458

Project Manager: Karen Horton
Editor-in-Chief: Sheri Snavely
Executive Managing Editor: Kathleen Schiaparelli
Assistant Managing Editor: Dinah Thong
Production Editor: Dana Dunn
Supplement Cover Management/Design: Paul Gourhan
Manufacturing Buyer: Ilene Kahn

© 2003 by Pearson Education, Inc.
Pearson Education, Inc.
Upper Saddle River, NJ 07458

The author and publisher of this book have used their best efforts in preparing this book. These efforts include the development, research, and testing of the theories and programs to determine their effectiveness. The author and publisher make no warranty of any kind, expressed or implied, with regard to these programs or the documentation contained in this book. The author and publisher shall not be liable in any event for incidental or consequential damages in connection with, or arising out of, the furnishing, performance, or use of these programs.

Printed in the United States of America

10 9 8 7 6 5 4 3 2 1

ISBN 0-13-100510-3

Pearson Education Ltd., *London*
Pearson Education Australia Pty. Ltd., *Sydney*
Pearson Education Singapore, Pte. Ltd.
Pearson Education North Asia Ltd., *Hong Kong*
Pearson Education Canada, Inc., *Toronto*
Pearson Educación de Mexico, S.A. de C.V.
Pearson Education—Japan, *Tokyo*
Pearson Education Malaysia, Pte. Ltd.
Pearson Education, *Upper Saddle River, New Jersey*

Contents

How to Increase Your Chances of Success in Genetics:

1. Attend Class
2. Read the Book
3. Do the Assigned Problems
4. Don't Cram
5. Study When There Are No Tests
6. Develop Confidence from Effort
7. Set Disciplined Study Goals
8. Learn Concepts
9. Be Careful with Old Exams
10. Don't "Second Guess" the Teacher

A first course in genetics can be a humbling experience for many students. The intent of this book is to help you understand introductory genetics as presented in the text **Genetics: A Molecular Perspective** by Klug and Cummings. It is possible that the lowest grades received in one's major, or even in one's undergraduate career, may be in genetics. It is not unusual for some students to become frustrated with their own inability to succeed in genetics. This frustration is felt by teachers as they field the following types of student comments.

"I studied all the material but failed your test."

"I must have a mental block to it. I just don't get it. I just don't understand what you are asking."

"Where did you get that question? I didn't see anything like that in the book or in my notes."

"This is the first test I have **ever** failed."

"I helped three of my friends last night and I got the lowest grade."

"I am getting a 'D' in your course and I have never received less than a 'B' in my whole life."

"I stayed up all night studying for your exam and I still failed."

Similar to Algebra

Think back to the first time you encountered "word problems" in your first algebra class. How many times did you ask yourself, your parents, or to your teacher the following classic question?

"I hate word problems, I just can't understand them, and why do I need to learn this anyway, I'll never use it?"

At that time you had two choices, drop out and be afraid of problem solving for the rest of your life (which unfortunately happens too often) or regroup, seek help, strip away distractions, and focus in on learning something new and powerful. Because you are taking genetics, you probably succeeded in algebra, perhaps with difficulty at first, and you will probably succeed in genetics.

In algebra you were forced to convert something real and dynamic (two trains leaving at different times from different stations at different speeds, when do they meet?) to a somewhat abstract formula which can be applied to an infinite number of similar problems. In genetics you will again learn something new. It will involve the conversion of something real and dynamic (genes, chromosomes, hereditary elements, gamete formation, gene splicing, and evolution) to an array of general concepts (similar to mathematical formulas) which will allow you to predict the outcome of an infinite number of presently known and yet to be discovered phenomena relating to the origin and maintenance of life.

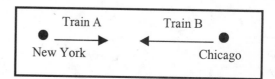

1

Students: Read This Section First!!!

Mental Pictures and Symbols

When working almost any "word" problem it is often helpful to make a simple drawing, which relates, in space, the primary participants. From that drawing one can often predict or estimate a likely outcome. A mathematical formula and its solution provide the precise outcome. To understand genetics it is often helpful to make drawings of the participants whether they be crosses (Aa X Aa), gametes (A or a), or the interactions of molecules (anticodon with codon).

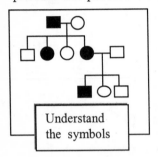

Understand the symbols

As with algebra, symbols used to represent a multitude of structures, movements, and interactions, are abstract, informative, and fundamental to understanding the discipline. It is the set of symbols and their interrelationships that comprise the concepts which make up the framework of genetics. Test questions and problems exemplify the concepts and may be completely unfamiliar to the student, nevertheless they refer directly to the basic concepts of genetics.

Attendance and Attention Are Mandatory

Many professors do not take attendance in lectures; therefore, it is likely that some students will opt to take a day off now and then. Unless those students are excellent readers and excellent students in general, continual absences will usually result in failure.

Remember how difficult it was to set up and understand the first algebra word problem on your own. It is likely that your ultimate source of understanding came from the course instructor. While using the text is important in your understanding of genetics, the teacher can walk you through the concepts and strategies much more efficiently than a text because a text is organized in a sequential manner. A good teacher can "cut and paste" an idea from here and there as needed.

To benefit from the wisdom of the instructor, the student must concentrate during the lecture session rather than sit, passively taking notes, assuming that the ideas can be figured out at a later date. Too often the student will not be able to relate to notes passively taken weeks before. In addition, the instructor will not be able to cover all the material in the text. Parts will be emphasized while other areas may be omitted entirely.

There is no magic formula for understanding genetics or any other discipline of significance. Learning anything, especially at the college level, requires time, patience, and confidence. First, a student must be willing to focus on the subject matter for an hour or so each day over the entire semester (quarter, trimester, *etc.*). Study time must be free of distractions and pressured by realistic goals.

The student must be patient and disciplined. It will be necessary to study when there are no assignments due and no tests looming.

> *Since it is the instructor who writes and grades the tests, who is in a better position to prepare the students for those tests?*

2

Students: Read This Section First!!!

The majority of successful students are willing to read the text ahead of the lecture material, spend time thinking about the concepts and examples, and work as many sample problems as possible. They study for a period of time, stop, then return to review the most difficult areas. They do not try to cram information into marathon study sessions a few nights before the examinations. While they may get away with that practice on occasion, more often than not, understanding the concepts in genetics requires more mature study habits and preparation.

Perhaps a Different Way of Thinking

Because the acquisition of problem-solving ability requires that students rely on new and important ways of seeing things rather than memorizing the book and notes, some students find the transition more difficult than others.  Some students are more able to deal in the abstract, concept-oriented framework than others. Students who have typically relied on "pure memory" for their success, will find a need to focus on concepts and problem-solving. They may struggle at first just as they may have struggled with the first word problem in algebra. But the reward for such struggle is intellectual growth. That's what college is supposed to stimulate. With such growth will come an increased ability to solve a variety of problems beyond genetics. Problem-solving is a process, a style, which can be applied to many disciplines. Few people are actually born with the touch of synthetic brilliance. Success comes from probing deeply in a few areas to see how problems are approached in a given discipline. Then, because problems are usually approached in a fairly consistent manner, a given problem-solving approach can often be applied to a variety of activities.

Read ahead. You have been told that it is important to read the assigned material before attending lectures. This allows you to make full use of the information provided in the lecture and to concentrate on those areas which are unclear in the readings. An opportunity is often provided for asking questions. Your questions will be received much more favorably if you can say that after reading the book and listening to the lecture a particular point is still unclear. It is very likely that your question will be quickly dealt with to your benefit and the benefit of others in the class.

> ### *Ask Questions and Don't Tune Out!*

How to Study

Genetics is a science which involves symbols (A, b, p), structures (chromosomes, ribosomes, plasmids), and processes (meiosis, replication, translation) which interact in a variety of ways. Models describe the manner in which hereditary units are made, how they function, and how they are transmitted from parent to offspring. Because many parts of the models interact both in time and space, genetics can not be viewed as a discipline filled with facts which should be memorized. Rather, one must be, or become, comfortable with seeking to understand not only the components of the models but also how the models work.

One can memorize the names and shapes of all the parts of an automobile engine, but without studying the interrelationships among the parts in time and space, one will have little understanding of the real nature of the engine. It takes time, work, and patience to see how an engine works and it will take time, work, and patience to understand genetics.

> ### *Time, Work, Patience*

3

Students: Read This Section First!!!

Don't cram. A successful tennis player doesn't learn to play tennis overnight; therefore, you can't expect to learn genetics under the pressure of night-long cramming. It will be necessary for you to develop and follow a realistic study schedule for genetics as well as the other courses you are taking. It is important that you focus your study periods into intensive, but short sessions each day throughout the entire semester (quarter, trimester). Because genetics tests often require you to think "on the spot" it is very important that you get a good night's sleep before each test. Avoid caffeine in the evening before the test because a clear, rested, well-prepared mind will be required.

> **Study when there are no tests**

Study goals. The instruction of genetics is often divided into large conceptual units. A test usually follows each unit. It will be necessary for you to study genetics on a routine basis long before each test. To do so, set specific study goals. Adhere to these goals and don't let examinations in one course interfere with the study goals of another course. Notice that each course being taken is handled in the same way - study ahead of time and don't cram.

Study each subject at least every other day --- especially when there are no tests!

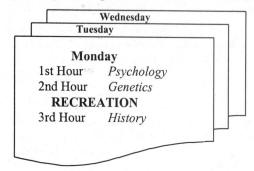

Develop a Realistic Monthly Schedule

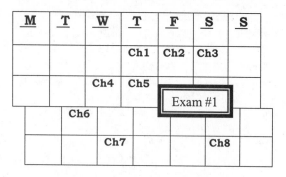

Develop a Plan for the Semester

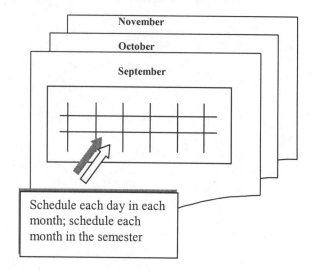

Schedule each day in each month; schedule each month in the semester

Work the assigned problems. The basic concepts of genetics are really quite straightforward but there are many examples which apply to these concepts. To help students adjust to the variety of examples and approaches to concepts, instructors often assign practice problems from the back of each chapter. If your instructor has assigned certain problems, finish working them *at least* one week before each examination. Before starting a set of problems, read the chapter carefully and consider the information presented in class.

Suggestions for working problems:

(1) Work the problem without looking at the answer.

(2) Check your answer in this book.

(3) If incorrect, work the problem again.

(4) If still incorrect, you don't understand the concept.

(5) Re-read your lecture notes and the text.

(6) Work the problem again.

(7) If you still don't understand the solution, mark it, and go to the next problem.

In your next study session, return to those problems which you have marked. Expect to make mistakes and learn from those mistakes. Sometimes what is difficult to see one day may be obvious the next day. If you are still having problems with a concept, schedule a meeting with your instructor. Usually the problem can be cleared up in a few minutes.

You will notice that in this book, I have presented the solution to each problem. I provide different ways of looking at some of the problems. Instructors often take a problem directly from those at the end of the chapters or they will modify an existing problem. Reversing the "direction" of a question is a common approach. Instead of giving characteristics of the parents and asking for characteristics of the offspring, the question may provide characteristics of the offspring and ask for particulars on the parents. Think as you work the problems.

Separate examples from concepts. As mentioned earlier, genetics boils down to a few (perhaps 15 to 20) basic concepts. However, there are many examples which apply to those concepts. Too often students have trouble separating examples from the concepts. Notice that in the "Sample Test" section in this book, I have made such separations clear. Examples allow you to picture, in concrete terms, various phenomena but they don't exemplify each phenomenon or concept in its entirety.

Be careful when using old examinations. Often it is customary for students to request or otherwise obtain old examinations from previous students. Such a practice is loaded with pitfalls. First, students often, albeit unconsciously, find themselves "second guessing" about questions on an upcoming examination. They forget that an examination usually only tests over a subset of the available information in a section. Therefore entire "conceptual areas" may be available which have not appeared on recent exams.

Often the reproductions of old examinations are of poor quality (having been copied and passed around repeatedly) and it is difficult to determine whether the answer provided is correct. In addition, if a question has the same general structure as one on a previous examination, but is modified, students often provide an answer for the "old" question rather than the one being asked. Granted, it is of value to see the format of each question and the general emphasis of previous examinations, but remember that each examination is potentially a new production capable of covering areas which have not been tested before. This is especially likely in a course such as genetics where the material changes very rapidly.

Students: Read This Section First!!!

Structure of this book

The intent of this book is to help you understand the concepts of genetics as given in the text and most likely in the lectures, then to apply these concepts to the solution of all problems and questions at the ends of the chapters. Rather than merely provide you with the solutions to the problems, I have tried to walk you through each component of each question so that you can see where information is obtained and how it can be applied in the solution. At the beginning of each chapter is a section which relates general concept areas to particular problems. This should help you practice certain conceptual areas as needed.

Vocabulary: Organization and listing of terms and concepts. Understanding the vocabulary of a discipline is essential to understanding the discipline. Throughout the text by Klug and Cummings you will find terms in bold print. Such terms generally refer to structures or substances, processes/methods, and concepts. I have separated these terms and *other important terms* into these categories.

> **Structures and Substances**
>
> **Processes/Methods**
>
> **Concepts**

Those terms or concepts which require special explanation or are more complex or intimately related to other terms are denoted with a code (**F2.1**, **F23.2**, *etc.*) which refers you to the figures immediately following each **Concepts** section of this book.

Use the listings as checklists to make certain that you understand the meaning of each term in each chapter. Also, by a given term's category, you can begin to understand whether it refers to a structure or substance, a process or method, or a more general concept. Notice that the various terms are not redefined. It is important that you use the Klug and Cummings text for the original definitions.

> **Understand the words and phrases of the discipline**

Concepts. In the section *Vocabulary: Organization and Listing of Terms and Concepts* you will find a section called *Concepts* after which there may be a simple sketch or two to help you focus a particular concept. Such sketches are oversimplifications and you should fill in the details by examining the textbook and the lecture notes.

Solved problems. Each of the problems at the end of each chapter is solved from a beginner's point of view. There are other features of this section. Many of the answers to the questions and problems will refer you to the Klug and Cummings (K/C) text. Be certain that you fully understand the solution to each of the questions suggested or assigned by your instructor.

Supplemental questions. A series of solved sample test questions supplement the questions provided in the text and help you determine your level of preparation. These sample test questions are located at the end of this book. Concepts relating to each question as well as common errors are presented in boxes before and after each answer.

> **Supplemental Questions**
> **Concepts**
> **Comprehensive Solution**
> **Common errors**

Chapter 1: Introduction to Genetics

Concept Areas	Corresponding Problems
Historical Aspects	1, 2, 3, 4
Basic Concepts of Genetics	5, 6
Investigative Approaches	7
Genetics and Society	8, 9, 10, 11, 12, 13

Vocabulary: Organization and Listing of Terms and Concepts

Historical

Prehistoric times

    domesticated animals

    cultivated plants

Greek Influence

    Hippocrates

    Aristotle

        epigenesis

        homunculus

    Wolff

    Schleiden and Schwann

    Harvey

        preformation

        spontaneous generation

        cell theory

    Galton (1883)

Evolution

    Darwin

        The Origin of Species

    natural selection

pangenesis

Lamarck

    acquired characteristics

    use and disuse

Genetics

    Mendel (1865)

    Rediscovery (1900)

        Correns

        de Vries

        Von Tschermak

    Bateson

    Watson and Crick (1953)

Structures and Substances

Center for heredity

    nucleus, nucleoid region

Genetic material

    DNA (deoxyribonucleic acid)

    RNA (ribonucleic acid)

        nitrogenous bases

    chromosomes

Chapter 1   Introduction to Genetics

chromatin fibers, centromere

Associated substances/structures

    amino acids (20)

    messenger RNA

    ribosome

    transfer RNA

    enzymes

    energy of activation

Processes/Methods

Mitosis

Meiosis

Transcription

Translation

Coding

    complementarity, hydrogen bonds

Transmission genetics

    pedigree analysis

Cytogenetics

    karyotypes

Population genetics

Molecular genetic analysis

    genomics

    bioinformatics

    recombinant DNA technology

Basic and applied research

Genetics and Society

eugenics (positive, negative)

euphenics

Agriculture

    "Green Revolution", Borlaug (1970)

Medicine

    genetic counseling

    recombinant DNA technology

    Human Genome Project

    human genetic engineering

Age of Genetics

Concepts

Geneology in Ireland

Historical aspects

Genetics

    variation, alleles

    genotype, phenotype

Chromosome

    diploid number (2n)

    polyploid

    haploid number

    homologous chromosomes

Genetic variation

    gene mutations

    chromosomal aberrations

Genetic information

    protein synthesis

Genetics and social issues

F1.1 Simple diagram of the relationships among major components of the *Trinity of Molecular Genetics*

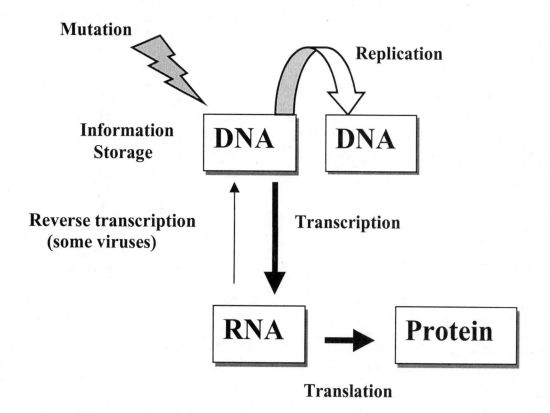

Solutions to Problems and Discussion Questions

1. Both were concerned with subjects of the reproduction, heredity, the origin of humans, and the shifting of interest from religious mythology to philosophical and scientific inquiries. Hippocrates argued that male semen is formed in various parts of the body (healthy or diseased) and transported through blood vessels to the testicles. Such "humors" carried the hereditary traits. Thus the theory of pangenesis was formed. Aristotle was critical of pangenesis because it did not explain the appearance of features which skipped generations. Aristotle suggested that semen contained a vital heat which could produce offspring in the form of the parents.

2. *Epigenesis* refers to the theory that organisms are derived from the assembly and reorganization of substances in the *egg* which eventually lead to the development of the adult. *Preformationism* is a 17th century theory, which states that the sex cells (eggs or sperm) contain miniature adults, called homunculi, which grow in size to become the adult. Each postulates a fundamental difference in the manner in which organisms develop from hereditary determiners.

3. Darwin was aware of the physical and physiological diversity of members within and among various species. He was aware that varieties of organisms could be developed through selective breeding (domestication), that a species is not a fixed entity, and that while certain groups of organisms could be hybridized, other groups could not. He was aware of conflicts between religious views and the fossil record. He understood geology, geography, and biology and that organisms tend to leave more offspring than the environment can support.

4. Their theory of natural selection proposed that more offspring are produced than can survive, and that in the competition for survival, those with favorable variations survive. Over many generations this will produce a change in the genetic make-up of populations if the favorable variations are inherited.

Darwin did not understand the nature of heredity and variation which led him to lean toward older theories of pangenesis and inheritance of acquired characteristics.

5. *Genes*, linear sequence of nucleotides, usually exert their influence by producing proteins through the process of transcription and translation. Genes are the functional units of heredity. They associate, sometimes with proteins, to form *chromosomes*. During the cell cycle, chromosomes and therefore genes, are duplicated by a variety of enzymes so that daughter cells inherit copies of the parental hereditary information. Genes of eukaryotes and prokaryotes are composed of DNA.

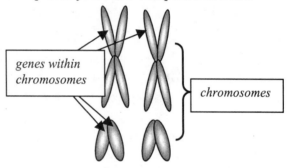

6. The "trinity" of molecular genetics refers to the relationships among DNA, RNA, and protein. The processes of *transcription* and *translation* are integral to understanding these relationships. See F1.1 above in this book.

7. *Transmission* genetics is the most classical approach in which the patterns of inheritance are studied through selective matings or the results of natural matings. Mendel observed results from precisely-defined matings and provided models based on transmission genetics. A second approach involves physical, traditionally microscopic (light and electron), examination of chromosomes. With the discovery of mitotic and meiotic processes, and the knowledge that genes are located on chromosomes, much interest centers on the *cytological investigation* of chromosomes.

Chapter 1   Introduction to Genetics

Molecular and biochemical analysis of the genetic material has recently evolved into one of the most exciting and rapidly growing subdisciplines of genetics. Originating in the early 1940s with studies of bacteria and viruses, much information as to the nature of gene expression, regulation, and replication has been provided. Recombinant DNA technology has had a significant impact in this area as well as others.

In *population genetics*, the interest is in the behavior of genes in groups of organisms (populations) often with an interest in the factors which change gene frequencies in time and space. Hence population geneticists are often interested in the process of evolution.

8. *Basic* research involves the study of the fundamental mechanisms of genetics as described in the answer to question #7 above. *Applied* research makes use of the applications in agriculture and medicine as described in the text.

9. Norman Borlaug applied Mendelian principles of hybridization and trait selection to the development of superior varieties of wheat. Such varieties are now grown in many countries, including Mexico, and have helped maintain the world supply of food. This change in worldwide agricultural food production has been called the "Green Revolution."

10. *Positive eugenics* encouraged parents displaying favorable characteristics to have large families while *negative eugenics* attempted to restrict reproduction for parents displaying unfavorable characteristics. *Euphenics* refers to medical genetic intervention designed to reduce the impact of defective genes on individuals.

11. Eugenic policies during this period were based on the mistaken premise that various "superior" and "inferior" traits are based on strict genetic control. Thus it was believed that by directing mating patterns, one could remove an unfavorable gene or genes from a given population. Sterilization was a method often taken to attempt to remove such genes. In fact, the role of the environment and basic principles of population genetics were largely ignored by those practicing eugenics.

12. In the last 40 years human transmission, cytological and molecular genetics have provided an understanding of many aspects of both plant and animal biology including development of pest resistant crops and identification of hazardous organisms in our food (*E. coli* for example). In addition, much has been learned about many human diseases. There is promise that a certain amount of human suffering will be minimized by the application of genetics to crop production (disease resistance, protein content, growth conditions) and medicine. Major medical areas of activity include genetic counseling, gene mapping and identification, disease diagnosis, and genetic engineering.

13. This question is open to many "answers" depending on the individual. Although it may be difficult to put yourself in this position, consider not only what your decision would be but also why.

Chapter 2: DNA Structure and Analysis

<u>Concept Areas</u>	<u>Corresponding Problems</u>
Central Dogma	1, 2, 8, 9, 19
Transformation	3, 4
Differential Labeling of Macromolecules	5, 6, 7
Genetic Variation	24, 33, 34
Model Building	15, 16, 28, 31, 35, 36
Nucleic Acid Structure	10, 11, 12, 13, 14, 17, 18, 20, 23, 27, 28, 29, 30, 32
Genomic Complexity	24
Analytical Methods	20, 21, 22, 24, 25, 26, 27, 30, 36, 37

Vocabulary: Organization and Listing of Terms and Concepts

Historical

1900-1944

    genetic material

        Miescher (nuclein)

        proteins

        nucleic acids

            tetranucleotide hypothesis

            base ratios, Erwin Chargaff

            transforming principle

                Avery *et al.* (1944)

                deoxycholate

            T2 bacteriophage (phage)

                Hershey and Chase (1952)

                    ^{32}P, ^{35}S

Watson and Crick (1953)

Franklin

Wilkins

Structures and Substances

Messenger RNA (mRNA)

Transfer RNA (tRNA)

Ribosomal RNA (rRNA)

Ribonuclease

Deoxyribonuclease

Lysozyme

Diplococcus

serotype

Protoplasts (spheroplasts)

φX174

Chapter 2 DNA Structure and Analysis

Ultraviolet light action spectrum

Insulin

Interferon

Human β-globin gene

RNA core

Coat protein

Qβ RNA replicase

Retrovirus

Reverse Transcriptase

Nucleic acids

    nucleotides

        nitrogenous base

            purines

                adenine

                guanine

            pyrimidines

                cytosine

                thymine

                uracil

        pentose sugar

            ribose

            deoxyribose

        phosphoric acid

nucleoside

    monophosphate

    diphosphate

    triphosphate

        adenosine triphosphate

        guanosine triphosphate

phosphodiester bond (5'-3')

inorganic phosphate

    hydrolysis

oligonucleotide

polynucleotide

base composition

    $A=T, G=C$

    $(A+G) = (C+T)$

    $(A+T)/(C+G) = $ variable

X-ray diffraction

antiparallel

0.34 nm (stacked bases) (3.4 Å)

3.4 nm (complete turn)

10 bases per turn

    10.4 bases per turn

2.0nm diameter (20 Å)

hydrogen bonds

    (A to T, G to C)

complementarity

Chapter 2 DNA Structure and Analysis

major and minor grooves

hydrophobic bases

hydrophilic backbone

A-DNA

B-DNA

C-DNA

D-DNA

E-DNA

Z-DNA

P- DNA

RNA

    ribosomal RNA (rRNA)

      ribosomes

    messenger RNA (mRNA)

      primary transcripts

    transfer RNA (tRNA)

    small nuclear RNA

    telomerase RNA

    antisense RNA

    biotin

    avidin (strepavidin)

      fluorescent label

Processes/Methods

Replication (F2.1)

    mitosis

    meiosis

Storage of information (F2.1)

Expression (F2.1)

    transcription

    translation

      central dogma

Variation (mutation) (F2.1)

Transformation

    Diplococcus pneumoniae

      Streptococcus pneumoniae

    virulent

    avirulent

    serotypes (II, III)

    smooth, rough

    heat-killed IIIS

    rough, extremely rough (ER)

      ribonuclease

      proteolytic enzymes

      deoxyribonuclease

    transfection

recombinant DNA research

   insulin

   interferon

   human β-globin gene

transgenic mice

   growth hormone

RNA as genetic material

   TMV (tobacco mosaic virus)

   Holmes ribgrass (HR)

     RNA core, coat protein

   Qβ phage

     Qβ RNA replicase

   retroviruses

     reverse transcription

Bonding

   sugar to purine

   sugar to pyrimidine

   nucleotide to nucleotide

Single crystal X-ray analysis

   Svedberg coefficient (S)

   Absorption of ultraviolet light (UV)

     254-260 nm

Sedimentation behavior

   gradient centrifugation

   sedimentation velocity

sedimentation equilibrium

Denaturation (melting)

   heat, chemical treatment

Spectrophotometry

Melting profile

   Melting temperature (T_m)

Hyperchromic effect

Renaturation (hybridization)

   DNA/DNA

   DNA/RNA

in situ hybridization

autoradiography

   FISH

   kinetics

   C_0t

   $C_0t_{1/2}$

   sequence complexity

electrophoresis

   polyacrylamide

   agarose

Concepts

Central Dogma

Characteristics of genetic material

Tetranucleotide hypothesis

Transformation

Differential labeling of macromolecules

Indirect evidence

    DNA content (*n, 2n*)

    mutagenesis

        action spectrum

        absorption spectrum

           260 nm, 280 nm

Direct Evidence

    recombinant DNA technology

Genetic variation

Model building

DNA double helix

    storage of genetic information

    information flow

    mutation

Genomic complexity

    reassociation kinetics

Separation strageties

Labeling strategies

F2.1 Illustration of relationships between DNA, its functions and related products.

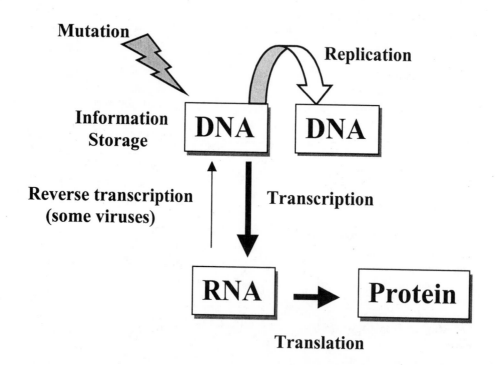

Solutions to Problems and Discussion Questions

1. *Replication* is that process which leads to the production of identical copies of the existing genetic information. Since daughter cells contain essentially exact copies (with some exceptions) of genetic information of the parent cell, and through the production and union of gametes, offspring contain copies (with variation) of parental genetic information, the genetic material must make copies of (replicate) itself. Replication is accomplished during the S phase of interphase.

The genetic material is capable of *expression* through the production of a phenotype. Through transcription and translation, proteins are produced which contribute to the phenotype of the organism. The genetic material must be stable enough to maintain information in "*storage*" from one cell to the next and one organism to the next. Because the genetic material is not "used up" in the processes of transcription and translation, genetic information can be stored and used constantly. Above, it was stated that the genetic material must be stable enough to store genetic information; however, variation through *mutation* provides the raw material for evolution. The genetic material is capable of a variety of changes, both at the chromosomal and nucleotide levels. See F2.1.

2. Prior to 1940 most of the interest in genetics centered on the transmission of similarity and variation from parents to offspring (transmission genetics). While some experiments examined the possible nature of the hereditary material, abundant knowledge of the structural and enzymatic properties of proteins generated a bias which worked to favor proteins as the hereditary substance. In addition, proteins were composed of as many as twenty different subunits (amino acids) thereby providing ample structural and functional variation for the multiple tasks which must be accomplished by the genetic

material. The tetranucleotide hypothesis (structure) provided insufficient variability to account for the diverse roles of the genetic material.

3. Griffith performed experiments with different strains of *Diplococcus pneumoniae* in which a heat-killed pathogen, when injected into a mouse with a live non-pathogenic strain, eventually led to the mouse's death. A summary of this experiment is provided in the K/C text. Examination of the dead mouse revealed living pathogenic bacteria. Griffith suggested that the heat-killed virulent (pathogenic) bacteria transformed the avirulent (non-pathogenic) strain into a virulent strain.

Avery and coworkers systematically searched for the transforming principle originating from the heat-killed pathogenic strain and determined it to be DNA. Taylor showed that transformed bacteria are capable of serving as donors of transforming DNA indicating that the process of transformation involves a stable alteration in the genetic material (DNA).

4. Transformation is dependent on a macromolecule (DNA) which can be extracted and purified from bacteria. During such purification however, other macromolecular species may contaminate the DNA. Specific degradative enzymes, proteases, RNase, and DNase were used to selectively eliminate components of the extract and, if transformation is concomitantly eliminated, then the eliminated fraction is the transforming principle. DNase eliminates DNA and transformation, therefore it must be the transforming principle.

5. Nucleic acids contain large amounts of phosphorus and no sulfur whereas proteins contain sulfur and no phosphorus. Therefore the radioisotopes ^{32}P and ^{35}S will selectively label nucleic acids and proteins, respectively.

The Hershey and Chase experiment is based on the premise that the substance injected into the bacterium is the substance responsible for producing the progeny phage and therefore must be the hereditary material. The experiment demonstrated that most of the ^{32}P -labeled material (DNA) was injected while the phage ghosts (protein coats) remained outside the bacterium. Therefore the nucleic acid must be the genetic material.

6. Actually phosphorus is found in approximately equal amounts in DNA and RNA. Therefore labeling with ^{32}P would "tag" both RNA and DNA. However, the T2 phage, in its mature state, contains very little if any RNA, therefore DNA would be interpreted as being the genetic material in T2 phage.

7. In theory, the general design would be appropriate in that some substance, if labeled, would show up in the progeny of transformed bacteria. However, since the amount of transforming DNA is extremely small compared to the genomic DNA of the recipient bacterium and its progeny, it would be technically difficult to assay for the labeled nucleic acid. In addition, it would be necessary to know that the small stretch of DNA which caused the genetic transformation was actually labeled. This in itself would be relatively easy using present-day recombinant DNA techniques; however, in earlier times, such specific labeling would have been difficult.

8. The early evidence would be considered indirect in that at no time was there an experiment, like transformation in bacteria, in which genetic information in one organism was transferred to another using DNA. Rather, by comparing DNA content in various cell types (sperm and somatic cells) and observing that the *action* and *absorption* spectra of ultraviolet light were correlated, DNA was considered to be the genetic material. This suggestion was supported by the fact that DNA was shown to be the genetic material in bacteria and some phage. Direct evidence for DNA being the genetic material comes from a variety of observations including gene transfer which has been facilitated by recombinant DNA techniques.

9. Some viruses contain a genetic material composed of RNA. The tobacco mosaic virus is composed of an RNA core and a protein coat. "Crosses" can be made in which the protein coat and RNA of TMV are interchanged with another strain (Holmes ribgrass). The source of the RNA determines the type of lesion, thus, RNA is the genetic material in these viruses. Retroviruses contain RNA as the genetic material and use an enzyme known as *reverse transcriptase* to produce DNA which can be integrated into the host chromosome. See F2.1.

10. The structure of deoxyadenylic acid is given below and in the K/C text. Linkages among the three components require the removal of water (H_2O).

11. The numbering of the carbons on the sugar is especially important (see diagram below). Examine the K/C text for the numbers on the carbons and nitrogens of the bases:

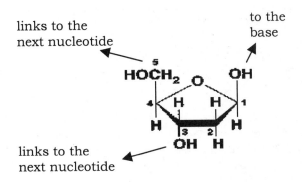

links to the
next nucleotide

to the
base

links to the
next nucleotide

12. Examine the structures of the bases in the K/C text. The other bases would be named as follows:

Guanine: 2-amino-6-oxypurine

Cytosine: 2-oxy-4-aminopyrimidine

Thymine: 2,4-dioxy-5-methylpyrimidine

Uracil: 2,4-dioxypyrimidine

13. Examine the K/C text for the format for this drawing. Note that the complementary strand must be drawn in the antiparallel orientation.

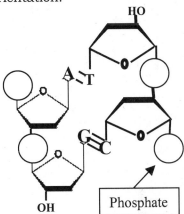

Phosphate

14. The following are characteristics of the Watson-Crick double-helix model for DNA:

The base composition is such that A=T, G=C and (A+G) = (C+T). Bases are stacked, 0.34 nm (3.4 Angstoms) apart, in a plectonic, antiparallel manner. There is one complete turn for each 3.4 nm which constitutes 10 bases per turn. Hydrogen bonds hold the two polynucleotide chains together, each being formed by phosphodiester linkages between the five-carbon sugars and the phosphates. There are two hydrogen bonds forming the A to T pair and three forming the G to C pair. The double helix exists as a twisted structure, approximately 20 Angstroms in diameter, with a topography of major and minor grooves. The hydrophobic bases are located in the center of the molecule while the hydrophilic phosphodiester backbone is on the outside.

15. In addition to creative "genius" and perseverance, model building skills, and the conviction that the structure would turn out to be "simple" and have a natural beauty in its simplicity, Watson and Crick employed the X-ray diffraction information of Franklin and Wilkins, and the base ratio information of Chargaff.

16. Because in double-stranded DNA, A=T and G=C (within limits of experimental error), the data presented would have indicated a lack of pairing of these bases in favor of a single-stranded structure or some other nonhydrogen-bonded structure.

Alternatively, from the data it would appear that A=G and T=C which would require purines to pair with purines and pyrimidines to pair with pyrimidines. In that case, the DNA would have contradicted the data from Franklin and Watkins which called for a constant diameter for the double-stranded structure.

Chapter 2 DNA Structure and Analysis

17. A covalent bond is a relatively strong bond which involves the sharing of electrons between two or more atoms. Hydrogen bonds, much weaker than covalent bonds, are formed as a result of

"electrostatic attraction between a covalently bonded hydrogen atom and an atom with an unshared electron pair. The hydrogen atom assumes a partial positive charge, while the unshared electron pair-characteristic of covalently bonded oxygen and nitrogen atoms- assumes a partial negative charge. These opposite charges are responsible for the weak chemical attraction."(Klug and Cummings)

Complementarity, responsible for the chemical attraction between adenine and thymine (uracil) and guanine and cytosine, is responsible for DNA and RNA assuming their double-stranded character. Complementarity is based on hydrogen bonding.

18. Three main differences between RNA and DNA are the following:

(1) uracil in RNA replaces thymine in DNA,

(2) ribose in RNA replaces deoxyribose in DNA, and

(3) RNA often occurs as both single- and partially double-stranded forms whereas DNA most often occurs in a double-stranded form.

19. While there are many types of RNA, the three main types described in this section are presented below:

ribosomal RNA: rRNA combines with proteins to form ribosomes which function to align mRNA and charged tRNA molecules during translation.

transfer RNA: tRNAs are involved in protein synthesis in that they represent a "link" between the codes in DNA (as reflected in mRNA) and the ordering of amino acids in proteins. Transfer RNAs are specific in that each species is attached to only one type of amino acid.

messenger RNA: the genetic code in DNA is transferred to the site of protein synthesis by a relatively short-lived molecule called messenger RNA. In eukaryotes, mRNA carries genetic information from the nucleus to the cytoplasm. It is the sequence of bases in mRNA which specifies the order of amino acids in proteins.

20. The nitrogenous bases of nucleic acids (nucleosides, nucleotides, and single- and double-stranded polynucleotides), absorb UV light maximally at wavelengths 254 to 260 nm. Using this phenomenon, one can often determine the presence and concentration of nucleic acids in a mixture. Since proteins absorb UV light maximally at 280 nm, this is a relatively simple way of dealing with mixtures of biologically important molecules.

UV absorption is greater in single-stranded molecules (hyperchromic shift) as compared to double-stranded structures, therefore one can easily determine, by applying denaturing conditions, whether a nucleic acid is in the single- or double-stranded form. In addition, A-T rich DNA denatures more readily than G-C rich DNA, therefore one can estimate base content by denaturation kinetics.

21. *Sedimentation velocity* centrifugation refers to an ultracentrifugation technique which monitors the velocity with which macromolecules move through a centrifugal field. Molecules move through the gradient on the basis of their mass and shape. If centrifuged long enough, such molecules will end up at the bottom of the tube.

Sedimentation equilibrium centrifugation is a technique which provides separation in a gradient on the basis of buoyant density. Macromolecules migrate through the gradient until they reach and subsequently remain at the point of equal density.

22. Guanine and cytosine are held together by three hydrogen bonds whereas adenine and thymine are held together by two. Because G-C base pairs are more compact, they are more dense than A-T pairs. The percentage of G-C pairs in DNA is thus proportional to the buoyant density of the molecule as illustrated in the K/C text.

23. Various treatments, heat, and certain chemical environments cause separation of the hydrogen bonds which hold together the complementary strands of DNA. Under these conditions, double-stranded DNA is changed to single-stranded DNA.

24. Carefully examine the K/C text. First understand the concept of molecular hybridization, then see that as the degree of strand uniqueness increases, the time required for reassociation increases. Repetitive sequences renature relatively quickly because the likelihood of complementary strands interacting increases.

For curve A in the problem, there is evidence for a rapidly renaturing species (repetitive) and a slowly renaturing species (unique). The fraction which reassociates faster than the *E. coli* DNA is highly repetitive and the last fraction (with the highest $C_ot_{1/2}$ value) contains primarily unique sequences. Fraction B contains mostly unique, relatively complex DNA.

25. *A hyperchromic effect* is the increased absorption of UV light as double-stranded DNA (or RNA for that matter) is converted to single-stranded DNA. As illustrated in the K/C text, the change in absorption is quite significant, with a structure of higher G-C content *melting* at a higher temperature than an A-T rich nucleic acid. If one monitors the UV absorption with a spectrophotometer during the melting process the hyperchromic shift can be observed. The T_m is the point on the profile (temperature) at which half (50%) of the sample is denatured.

26. Because G-C base pairs are formed with three hydrogen bonds while A-T base pairs by two such bonds, it takes more energy (higher temperature) to separate G-C pairs.

27. The reassociation of separate complementary strands of a nucleic acid, either DNA or RNA, is based on hydrogen bonds forming between A-T (or U) and G-C.

28. In one sentence of Watson and Crick's paper in *Nature*, they state,

> "It has not escaped our notice that the specific pairing we have postulated immediately suggests a possible copying mechanism for the genetic material."

The model itself indicates that unwinding of the helix and separation of the double-stranded structure into two single strands immediately exposes the specific hydrogen bonds through which new bases are brought into place.

29.

(1) As shown, the extra phosphate is not normally expected.

(2) In the adenine ring, a nitrogen is at position 8 rather than position 9.

(3) The bond from the C-1' to the sugar should form with the N at position 9 (N-9) of the adenine.

(4) The dinucleotide is a "deoxy" form, therefore each C-2' should not have a hydroxyl group. Notice the hydroxyl group at C-2' on the sugar of the adenylic acid.

(5) At the C-5 position on the thymine residue, there should be a methyl group.

(6) At the C-5 position on the thymidylic acid, there is an extra OH group.

30. As provided in the text, a direct proportionality between $C_0t_{1/2}$ and the number of base pairs exists under certain conditions.

The ratios for MS-2 would be as follows:

$0.5/10^5 = 0.001/X$

or

$X/0.001 = 10^5/0.5$

$X = (0.001)(10^5)/0.5$

$X = 200$ base pairs

The ratios for *E. coli* would be as follows:

$0.5/10^5 = 10.0/X$ or

$X/10.0 = 10^5/0.5$

$X = (10.0)(10^5)/0.5$

$X = 2 \times 10^6$ base pairs

31.

(i) The X-ray diffraction studies would indicate a helical structure, for it is on the basis of such data that a helical pattern is suggested. The fact that it is irregular may indicate different diameters (base pairings), additional strands in the helix, kinking or bending.

(ii) The hyperchromic shift would indicate considerable hydrogen bonding, possibly caused by base pairing.

(iii) Such data may suggest irregular base pairing in which purines bind purines (all the bases presented are purines) thus giving the atypical dimensions.

(iv) Because of the presence of ribose, the molecule may show more flexibility, kinking, and/or folding.

While there are several situations possible for this model, the phosphates are still likely to be far apart (on the outside) because of their strong like charges. Hydrogen bonding probably exists on the inside of the molecule and there is probably considerable flexibility, kinking, and/or bending.

32. Left side (a) = left, right side (b) = right.

33. Since cytosine pairs with guanine and uracil pairs with adenine, the result would be a base substitution of G:C to A:T after rounds of replication.

34. Under this condition, the hydrolyzed 5-methyl cytosine becomes thymine.

35. Without knowing the exact bonding characteristics of hypoxanthine or xanthine it may be difficult to predict the likelihood of each pairing type. It is likely that both are of the same class (purine or pyrimidine) because the names of the molecules indicate a similarity. In addition, the diameter of the structure is constant which, under the model to follow, would be expected. In fact, hypoxanthine and xanthine are both purines.

Because there are equal amounts of A, T and H, one could suggest that they are hydrogen bonded to each other; the same may be said for C, G, and X. Given the molar equivalence of erythrose and phosphate, an alternating sugar-phosphate-sugar backbone as in "earth-type" DNA would be acceptable. A model of a triple helix would be acceptable, since the diameter is constant. Given the chemical similarities to "earth-type" DNA it is probable that the unique creature's DNA follows the same structural plan.

36. (a) Heat application would yield a hyprochromic if the DNA is double-stranded. One could also get a rough estimation of the GC content from the kinetics of denaturation and the degree of sequence complexity from comparative renaturation studies.

(b) Determination of base content by hydrolysis and chromatography could be used for comparative purposes and could also provide evidence as to the strandedness of the DNA.

(c) Antibodies for Z-DNA could be used to determine the degree of left-handed structures, if present.

(d) Sequencing the DNA from both viruses would indicate sequence homology. In addition, through various electronic searches readily available on the Internet (wed site: *blast @ ncbi . nlm . nih . gov*, for example) one could determine whether similar sequences exist in other viruses or in other organisms.

37. The way the question is stated suggests that DNA which is separated electrophoretically is of the same shape (long rod). In fact, DNA can exist in a variety of shapes as seen in supercoiled plasmids, relaxed (nicked) plasmids, and linear molecules. Size comparisons with DNA must be such that linear molecules are compared with linear molecules and supercoiled with supercoiled, *etc*. In comparing DNA migration to RNA, even though RNA molecules have the same charge to mass ratios, they also exist in a variety of shapes. Complementary intra-strand base pairing can make more compact structures compared to the more relaxed, open conformation. For electrophoretic size comparisons, RNA molecules must be denatured to eliminate secondary structural variables.

Chapter 3: DNA Replication and Recombination

Concept Areas	Corresponding Problems
Replication	1, 2, 3, 4, 5, 6, 7, 12, 13, 14, 18, 21, 22, 23, 24, 28, 29, 31, 32, 33
Nearest Neighbor Analysis	8, 9, 10, 11, 12, 23, 30
Enzymology	6, 15, 16, 17, 19, 20, 27, 28, 32
Conditional Mutations	25
Gene Conversion	26

Vocabulary: Organization and Listing of Terms and Concepts

Structures and Substances

DNA polymerase I

Spleen phosphodiesterase (F3.1)

5'-nucleotides (F3.2)

3'-nucleotides (F3.2)

Phage φX174

    + strand, - strand

    replicative form (RF)

    5-bromouracil (BU)

DNA ligase (polynucleotide joining enzyme)

DNA polymerase II

DNA polymerase III

    primer

    exonuclease

    holoenzyme

    subunits

dimer, γ complex

replisome

Helicases, *dna*A, *dna*B, *dna*C

Single-stranded DNA binding proteins

DNA gyrase (topoisomerase)

RNA primer

    primase

    free 3' hydroxyl group

DNA ligase

    ligase deficient mutant

ori C, *ter*

    9mer, 13mer

    β-subunit clamp

Chapter 3 DNA Replication and Recombination

Eukaryotic DNA polymerases

    six forms

    multiple replicons

Antonomously replicating sequence (ARS)

Origin replication complex (ORC)

Pre-replication complex

Nucleosome

Telomerase

    Tetrahymena

    TTGGGG

    hairpin loop

    catalytic ribonucleoprotein

Endonuclease

Heteroduplex DNA molecules

    Holliday structures

    chi form

    recombinant duplexes

    *rec*A, *rec*B, *rec*C, *rec*D

Processes/Methods

Replication of DNA

    semiconservative

    conservative

    dispersive

Meselson and Stahl - 1958

    E. coli

    equilibrium sedimentation

    $^{15}NH_4Cl$, $^{14}NH_4Cl$

Taylor, Woods, and Hughes - 1957

    Vicia faba

    ^{3}H-thymidine

    autoradiography

    colchicine

    sister chromatid exchanges

bidirectional (*vs.* unidirectional)

    origin of replication, *ori*

    termination, *ter*

    replicon

    replication fork

    continuous, discontinuous

    leading strand

    lagging strand

    Okazaki fragments

Synthesis of DNA *in vitro*

    Kornberg - 1957

    reaction mixture

    chain elongation

    fidelity

Chapter 3 DNA Replication and Recombination

base comparisons (template/product)

nearest neighbor frequency test (F10.1)

    (see problem #29 for explanation)

    spleen phosphodiesterase

biologically active DNA

transfection of *E. coli*

faithful copying

Processivity

Polymerase switching

Exonuclease proofreading

Conditional mutation

    temperature sensitive

Genetic recombination

    homologous recombination

    single-stranded nick

    endonuclease and ligation

Gene conversion

    Neurospora

    nonreciprocal

Concepts

Replication

    semiconservative

    antiparallel

    continuous, discontinuous

    conservative (F3.4)

    dispersive (F3.4)

  Biological activity

  Nearest neighbor analysis

Repair

Proofreading

Eukaryotic DNA replication

Telomere replication

Conditional mutants (F3.3)

Genetic recombinaiton

Gene conversion

F3.1 Illustration of the mode of action of spleen phosphodiesterase.

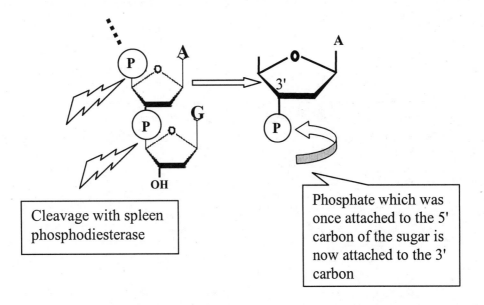

Cleavage with spleen phosphodiesterase

Phosphate which was once attached to the 5' carbon of the sugar is now attached to the 3' carbon

F3.2 Shorthand structures for 3' and 5' nucleotides.

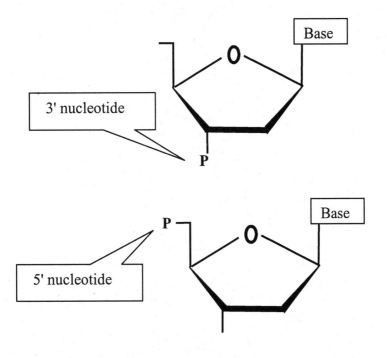

3' nucleotide

5' nucleotide

F3.3 Illustration of the influence of a conditional mutation on protein structure and function

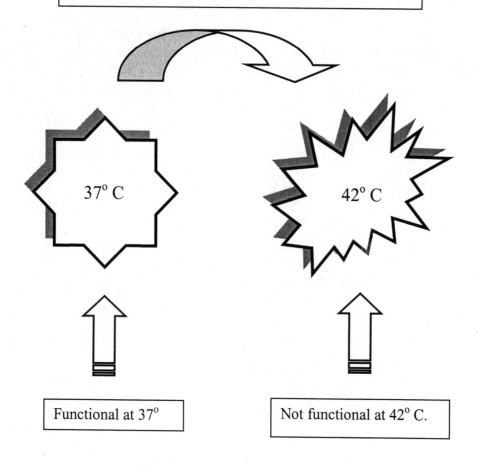

Changes in the environment of the protein may cause conformational changes in the protein to alter function of that protein.

37° C

42° C

Functional at 37°

Not functional at 42° C.

F3.4 Figure relating to question #4 in the problems section depicts labeling pattern under Conservative and Dispersive replication patterns.

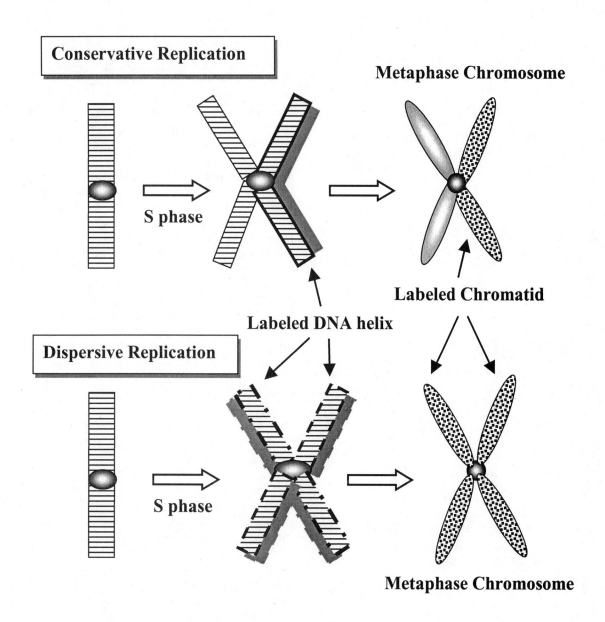

Solutions to Problems and Discussion Questions

1. The differences among the three models of DNA replication relate to the manner in which the new strands of DNA are oriented as daughter DNA molecules are produced.

Conservative: In the conservative scheme, the original double helix remains as a complete unit and the new DNA double helix is produced as a single unit. The old DNA is completely conserved.

Semiconservative: Each daughter strand is composed of one old DNA strand and one new DNA strand. Separation of hydrogen bonds is required.

Dispersive: In the dispersive scheme, the original DNA strand is broken into pieces and the new DNA in the daughter strand is interspersed among the old pieces. Separation of the individual covalent, phosphodiester bonds is required for this mode of replication.

2. The Meselson and Stahl experiment has the following components. By labeling the pool of nitrogenous bases of the DNA of *E. coli* with a heavy isotope ^{15}N, it would be possible to "follow" the "old" DNA. This is accomplished by growing the cells for many generations in medium containing ^{15}N. Cells were transferred to ^{14}N medium so that "new" DNA could be detected. A comparison of the density of DNA samples at various times in the experiment (initial ^{15}N culture, and subsequent cultures grown in the ^{14}N medium) showed that after one round of replication in the ^{14}N medium, the DNA was half as dense (intermediate) as the DNA from bacteria grown only in the ^{15}N medium. In a sample taken after two rounds of replication in the ^{14}N medium, half of the DNA was of the intermediate density and the other half was as dense as DNA containing only ^{14}N DNA.

3. Under a conservative scheme the first round of replication in ^{14}N medium produces one dense double helix and one "light" double helix in contrast to the intermediate density of the DNA in the semiconservative mode. Therefore, after one round or replication in the ^{14}N medium the conservative scheme can be ruled out.

After one round of replication in ^{14}N under a dispersive model, the DNA is of intermediate density, just as it is in the semiconservative model. However, in the next round of replication in ^{14}N medium, the density of the DNA is between the intermediate and "light" densities.

4. Refer to the text for an illustration of the labeling of *Vicia* chromosomes under a Taylor, Woods, and Hughes experimental design. Notice that only those cells which pass through the S phase in the presence of the ^3H-thymidine are labeled and that each double helix (per chromatid) is "half-labeled." See Figure F3.4 in this book for a graphic description of this Conservative and Dispersive replication patterns.

(a) Under a conservative scheme all of the newly labeled DNA will go to one sister chromatid, while the other sister chromatid will remain unlabeled. In contrast to a semiconservative scheme, the first replicative round would produce one sister chromatid which has label on both strands of the double helix. (See F3.4 above)

(b) Under a dispersive scheme all of the newly labeled DNA will be interspersed with unlabeled DNA. Because these preparations (metaphase chromosomes) are highly coiled and condensed structures derived from the "spread out" form at interphase (which includes the S phase) it is impossible to detect the areas where label is not found. Rather, both sister chromatids would appear as evenly labeled structures. (See F3.4 above)

5. Because the semiconservative scheme predicts that *half* of the DNA in each daughter double helix is labeled, it would be difficult to envision a scheme where three strands are replicated in such a semiconservative manner. It would seem that either the conservative or dispersive scheme would fit more appropriately. To examine the nature of replication, one could devise an experiment similar to that of Meselson and Stahl or Taylor, Woods, and Hughes.

6. The *in vitro* replication requires a DNA template, a divalent cation (Mg⁺⁺), and all four of the deoxyribonucleoside triphosphates: dATP, dCTP, dTTP, and dGTP. The lower case "d" refers to the deoxyribose sugar.

7. Prior to the development of highly efficient methods of enzyme isolation, large cultures, containing large numbers of bacterial cells, were needed to yield even small quantities of enzymes.

8. Two general analytical approaches showed that the products of DNA polymerase I were probably copies of the template DNA. Because *base composition* can be similar without reflecting sequence similarity, the least stringent test was the comparison of base composition. By comparing *nearest neighbor frequencies*, Kornberg determined that there is a very high likelihood that the product was of the same base sequence as the template.

9. Because base composition can be similar without reflecting sequence similarity, the least stringent test was the comparison of base composition. By comparing nearest neighbor frequencies Kornberg determined that there is a very high likelihood that the product was of the same base sequence as the template. However, since nearest neighbor analysis relates "neighbor frequency" and not base sequence, there is a chance that sequence differences are not detected.

10. The *in vitro* rate of DNA synthesis using DNA polymerase I is slow, being more effective at replicating single-stranded DNA than double-stranded DNA. In addition, it is capable of degrading as well as synthesizing DNA. Such degradation suggested that it functioned as a repair enzyme. In addition, DeLucia and Cairns discovered a strain of *E. coli* (*pol*A1) which still replicated its DNA but was deficient in DNA polymerase I activity.

11. See F3.1 and Problem #30 below to get started. The overall plan of the *nearest neighbor experiment* is to determine the frequency of neighbors of a given base, say adenine, in both the template DNA and the product DNA. Should the frequency of neighbors be similar, then the sequence of bases is likely to be the same. This argument is strengthened if the frequencies of neighbors of the other bases (G, C, and T) are also similar in the template and primer.

The procedure is to introduce labeled phosphate by use of labeled (innermost phosphate) 5'-nucleotides as shown in the K/C text. In the example, cytidine has a ³²P attached to the C-5' atom. Spleen phosphodiesterase cleaves the polymer at the C-5' position (between the C-5' and the phosphate) thereby transferring the labeled phosphate to the C-3' atom of its 5' neighbor.

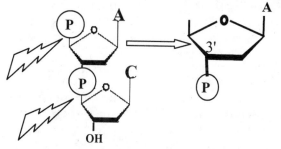

Chromatographic separation of the resulting 3'-nucleotides, followed by determination of the radioactivity of each, provides an estimate of the frequency of which each base is a neighbor of cytosine. By repeating this procedure using other bases and other templates, one can be somewhat confident that the template is similar in base sequence to the *in vitro* product.

12. As stated in the text, *biologically active* DNA implies that the DNA is capable of supporting typical metabolic activities of the cell or organism and is capable of faithful reproduction.

13. φX174 is a well-studied single-stranded virus (phage) which can be easily isolated. It has a relatively small DNA genome (5500 nucleotides) which, if mutated, usually alters its reproductive cycle.

14. As shown in the K/C text, DNA polymerase I and DNA ligase are used to synthesize and label an RF duplex. DNase is used to nick one of the two strands. The heavy (^{32}P/BU-containing) DNA is isolated by denaturation and centrifugation and DNA polymerase and DNA ligase are used to make a synthetic complementary strand. When isolated, this synthetic strand is capable of transfecting *E. coli* protoplasts, from which new φX174 phages are produced.

15. The *pol*AI mutation was instrumental in demonstrating that DNA polymerase I activity was not necessary for the *in vivo* replication of the *E. coli* chromosome. Such an observation opened the door for the discovery of other enzymes involved in DNA replication.

16. All three enzymes share several common properties. First, none can *initiate* DNA synthesis on a template but all can *elongate* an existing DNA strand assuming there is a template strand as shown in the figure below. Polymerization of nucleotides occurs in the 5' to 3' direction where each 5' phosphate is added to the 3' end of the growing polynucleotide.

5'   3'

synthesis can initiate here

5'

All three enzymes are large complex proteins with a molecular weight in excess of 100,000 daltons and each has 3' to 5' exonuclease activity. Refer to K/C text.

DNA polymerase I:
    exonuclease activity
    present in large amounts
    relatively stable
    removal of RNA primer
DNA polymerase II:
    possibly involved in repair function
DNA polymerase III:
    exonuclease activity
    essential for replication
    complex molecule

17. Refer to K/C text for a listing of the components of DNA polymerase III. The active form of the enzyme is called the holoenzyme. The region responsible for actual polymerization is called the "core" portion.

18. Given a stretch of double-stranded DNA, one could initiate synthesis at a given point and either replicate strands in one direction only (unidirectional) or in both directions (bidirectional) as shown below. Notice that in the K/C text the synthesis of complementary strands occurs in a *continuous* 5'>3' mode on the leading strand in the direction of the replication fork, and in a *discontinuous* 5'>3' mode on the lagging strand opposite the direction of the replication fork. Such discontinuous replication forms Okazaki fragments.

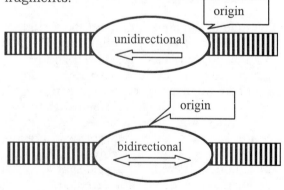

19. *Helicase,* dnaA and *single-stranded* DNA *binding* proteins initially unwind, open, and stabilize DNA at the initiation point. DNA *gyrase,* a DNA topoisomerase, relieves supercoiling generated by helix unwinding. This process involves breaking both strands of the DNA helix.

20. *Okazaki fragments* are relatively short (1000 to 2000 bases in prokaryotes) DNA fragments which are synthesized in a discontinuous fashion on the lagging strand during DNA replication. Such fragments appear to be necessary because template DNA is not available for 5'>3' synthesis until some degree of continuous DNA synthesis occurs on the leading strand in the direction of the replication fork. The isolation of such fragments provides support for the scheme of replication shown in the K/C text. DNA *ligase* is required to form phosphodiester linkages in gaps which are generated when DNA polymerase I removes RNA primer and meets newly synthesized DNA ahead of it. Notice in the K/C text the discontinuous DNA strands are ligated together into a single continuous strand. *Primer* RNA is formed by RNA primase to serve as an initiation point for the production of DNA strands on a DNA template. None of the DNA polymerases are capable of initiating synthesis without a free 3' hydroxyl group. The primer RNA provides that group and thus can be used by DNA polymerase III.

21. The synthesis of DNA is thought to follow the pattern described in the K/C text. The model involves opening and stabilization of the DNA helix, priming DNA with synthesis with RNA primer, movement of replication forks in both directions which includes elongation of RNA primers in continuous and discontinuous 5'>3' modes and their removal by the exonucleolytic activity of DNA polymerase I. Okazaki fragments generated in the replicative process are joined together with DNA ligase. DNA gyrase relieves supercoils generated by DNA unwinding.

22. Eukaryotic DNA is replicated in a manner which is very similar to that of *E. coli.* Synthesis is bidirectional, continuous on one strand and discontinuous on the other, and the requirements of synthesis (four deoxyribonucleoside triphosphates, divalent cation, template, and primer) are the same. Okazaki fragments of eukaryotes are about one-tenth the size of those in bacteria.

Because there is a much greater amount of DNA to be replicated and DNA replication is slower, there are multiple initiation sites for replication in eukaryotes (and increased DNA polymerase per cell) in contrast to the single replication origin in prokaryotes. Replication occurs at different sites during different intervals of the S phase. The proposed functions of four DNA polymerases are described in the text.

23. Even though the base composition between two species may be *similar,* sequences can vary considerably.

24. (a) In *E. coli,* 100kb are added to each growing chain per minute. Therefore the chain should be about 4,000,000bp.

(b) Given $(4 \times 10^6$ bp$) \times 0.34$nm/bp =

1.36×10^6nm or 1.3mm

25. (a) No repair from DNA polymerase I and/or DNA polymerase III. **(b)** No DNA ligase activity. **(c)** No primase activity. **(d)** Only DNA polymerase I activity. **e)** No DNA gyrase activity.

26. *Gene conversion* is likely to be a consequence of genetic recombination in which nonreciprocal recombination yields products in which it appears that one allele is "converted" to another. Gene conversion is now considered a result of heteroduplex formation which is accompanied by mismatched bases. When these mismatches are corrected, the "conversion" occurs.

27. (a) Because DNA polymerase III is essential for DNA chain elongation, it is necessary for replication of the *E. coli* chromosome. Thus strains which are mutant for this enzyme must contain conditional mutations. **(b)** The 3' - 5' exonuclease activity is involved in proofreading. Thus proofreading would be hampered in such mutant strains and a higher than expected mutation rate would occur.

28. Telomerase activity is present in germ line tissue to maintain telomere length from one generation to the next. In other words, telomeres can not shorten indefinitely without eventually eroding genetic information.

29. Since synthesis is bidirectional, one can multiply the rate of synthesis by two to come up with a figure of 18,000 bases replicated per five minutes (30bases/second X 300 seconds). Dividing 1.6×10^8 by 1.8×10^4 gives 0.88×10^4 or about 8,800 replication sites.

30.

Initial Labeled Base	Labeled Base after Spleen Phosphodiesterase Digestion	
	ANTIPARALLEL	PARALLEL
G	A,T	C,T
C	G,A,G	G,A
T	C,T, G	C,T,A,G
A	T,C,A,T	T,C,A,G

One can determine which model occurs in nature by comparing the pattern in which the labeled phosphate is shifted following spleen phosphodiesterase digestion. Focus your attention on the antiparallel model and notice that the frequency which "C" (for example) is the 5' neighbor of "G" is not necessarily the same as the frequency which "G" is the 5' neighbor of "C." However, in the parallel model (b), the frequency which "C" is the 5' neighbor of "G" is the same as the frequency which "G" is the 5' neighbor of "C." By examining such "digestion frequencies" it can be determined that DNA exists in the opposite polarity.

31. If replication is conservative, the first autoradiographs (see metaphase I in the K/C text) would have label distributed only on one side (chromatid) of the metaphase chromosome as shown below.

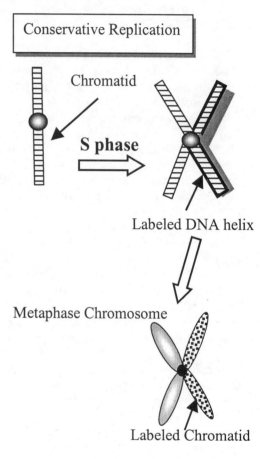

32.

(a) DNA polymerase would catalyze a bond between the 5' end of the last nucleotide added and the 3' end of the incoming nucleotide. In this reaction, the energy would be provided by the cleavage of the gamma- and beta- phosphates of the last nucleotide added to the chain rather than of the incoming nucleotide.

(b) If DNA polymerase removed a base, it would not be able to add any more bases to the chain because the penultimate base would have a monophosphate rather than a triphosphate and there would be no source of energy for the polymerization reaction.

33. If the DNA contained parallel strands in the double helix and the polymerase would be able to accommodate such parallel strands, there would be continuous synthesis and no Okazaki fragments. The telomere problem would only be at one end. Several other possibilities exist. If the DNA strands were replicated as complete single strands, the synthesis could begin at the opposite free ends. In addition, if the DNA existed only as a single strand, the same results would occur.

Chapter 4: Chromosome Structure and DNA Sequence Organization

Concept Areas	Corresponding Problems
Viral and Bacterial Chromosomes	1, 2, 10, 11
Organization of DNA in Chromatin	6, 7, 8, 9, 12
Organization of the Eukaryotic Genome	3, 4, 5, 6, 13

Vocabulary: Organization and Listing of Terms and Concepts

Structures and Substances

Viral chromosomes

DNA, RNA

double-stranded

single-stranded

often circular

protein coat

φX174

polyoma

lambda (λ)

Bacterial chromosomes

DNA

double-stranded

nucleoid

E. coli

circular

DNA-binding proteins

HU, H

topoisomer

topoisomerase

Mitochondrial DNA (mtDNA)

plant

animal

coding (mtDNA)

rRNAs

tRNAs

respiratory components

coding (nuclear)

imported products

Chloroplast DNA (cpDNA)

circular

double-stranded

different than nuclear DNA

Chapter 4 Chromosome Structure and DNA Sequence Organization

coding (cpDNA)

  rRNAs

  tRNAs

  ribulose-1-5-bisphosphate carboxylase

Eukaryotic chromosomes

  chromatin

  mitotic chromosomes

    condensed chromatin

      folded fiber

Chromatin

  nucleoprotein

  histones

    amino acid composition

    tetramers

      nucleosome core particle

  nonhistones

  micrococcal nuclease

    nucleosomes

    linker DNA

      histone H1

    solenoid

Heterochromatin, euchromatin

  telomere

  satellite and repetitive DNA

  pseudogene

Processes/Methods

Supercoiling

  supercoil

  linking number

    energetically relaxed

    energetically strained

Uniparental mode of inheritance

Semiconservative replication

Importing of nuclear-coded gene products

Autoradiography

Folded-fiber (eukaryotic chromosome)

    coiling-twisting-condensing

Chromatin remodeling

  methylation

  phosphorylation

Heterochromatin

  few genes

  late replicating

  position effect

In situ hybridization

Chromosome banding

  C-banding

  Q-banding

  G-banding

  R-banding

Chapter 4 Chromosome Structure and DNA Sequence Organization

Repetitive DNA

   non-coding sequences

      satellite DNA

      highly repetitive DNA

   centromeric DNA (CEN)

      alphoid family

   telomeric DNA sequences

      telomeric-associated sequences

      telomerase

   moderately repetitive DNA

   short interspersed elements (SINES)

      Alu family

   long interspersed elements (LINES)

   variable number tandem repeats (VNTR)

   DNA fingerprinting

minisatellites

microsatellites

repetitive transposed sequences

moderately repetitive multicopy genes

Introns, exons

   heterogeneous nuclear RNA (hnRNA)

Concepts

Variety of DNA conformations

Evolution of cellular organelles

   endosymbiont theory

Heterochromatin/euchromatin

Centromeric structure

Repetitive DNA

Chapter 4 Chromosome Structure and DNA Sequence Organization

Solutions to Problems and Discussion Questions

1. Bacteriophage λ has a linear, double-stranded DNA while in the phage coat and upon infection closes to form a circular chromosome. It contains about 50kb. T2 phage also has a linear, double-stranded DNA chromosome; less than 200kb. *E. coli* has a circular, double-stranded DNA chromosome of about 4.2×10^3kb. Both intact phage are about 1/150 the size of *E.coli*. Since phage are obligate parasites of bacteria, they are dependent on their hosts for the manufacture of materials for their replication. Bacteria contain all genetic information for metabolism, replication, and *de novo* synthesis of numerous life-supporting materials. Phage on the other hand contain relatively few genes; namely, those needed to adsorb, inject, and produce progeny using primarily bacterial materials.

2. By having a circular chromosome, no free ends present the problem of linear chromosomes, namely complete replication of terminal sequences.

3. Polytene chromosomes are formed from numerous DNA replications, pairing of homologues, and absence of strand separation or cytoplasmic division. Each chromosome contains about 1000-5000 DNA strands in parallel register. They appear in specific tissues, such as salivary glands, of many dipterans like *Drosophila*. They appear as comparatively long, wide, fibers with sharp light and dark sections (bands) along their length. Such bands (chromomeres) are useful in chromosome identification, *etc.*

4. Puffs represent active genes as evidenced by staining and uptake of labeled RNA precursors as assayed by autoradiography.

5. Lampbrush chromosomes are typically present in vertebrate oocytes and are so named because of their similar appearance to brushes used to clean kerosene lamp chimneys in the 19th century. They are also found in spermatocytes of some insects. They are found as diplotene stage structures and are active uncoiled versions of condensed meiotic chromosomes. Lampbrush chromosomes are typically viewed using light and electron microscopy.

6. While greater DNA content per cell is associated with eukaryotes, one can not universally equate genomic size with an increase in organismic complexity. There are numerous examples where DNA content per cell varies considerably among closely related species. Because of the diverse cell types of multicellular eukaryotes, a variety of gene products is required, which may be related to the increase in DNA content per cell. In addition, the advantage of diploidy automatically increases DNA content per cell. However, seeing the question in another way, it is likely that a much higher *percentage* of the genome of a prokaryote is actually involved in phenotype production than in a eukaryote.

Eukaryotes have evolved the capacity to obtain and maintain what appears to be large amounts of "extra" perhaps "junk" DNA. This concept will be examined in subsequent chapters of the text. Prokaryotes on the other hand, with their relatively short life cycle, are extremely efficient in their accumulation and use of their genome.

Given the larger amount of DNA per cell and the requirement that the DNA be partitioned in an orderly fashion to daughter cells during cell division, certain mechanisms and structures (mitosis, nucleosomes, centromeres, *etc.*) have evolved for packaging and distributing the DNA. In addition, the genome is divided into separate entities (chromosomes) to perhaps facilitate the partitioning process in mitosis and meiosis.

Chapter 4 Chromosome Structure and DNA Sequence Organization

7. Digestion of chromatin with endonucleases, such as micrococcal nuclease, gives DNA fragments of approximately 200 base pairs or multiples of such. X-ray diffraction data indicated a regular spacing of DNA in chromatin. Regularly spaced bead-like structures (nucleosomes) were identified by electron microscopy. Nucleosomes are octomeric structures of two molecules of each histone (H2A, H2B, H3, and H4) except H1. Between the nucleosomes and complexed with linker DNA is histone H1. A 146-base pair sequence of DNA wraps around the nucleosome and as chromosome condensation occurs a 300-Å fiber is formed. It appears to be composed of 5 or 6 nucleosomes coiled together. Such a structure is called a solenoid.

8. *Heterochromatin* is chromosomal material which stains deeply and remains condensed when other parts of chromosomes, euchromatin, are otherwise pale and decondensed. Heterochromatic regions replicate late in S phase and are relatively inactive in a genetic sense because there are few genes present or if they are present, they are repressed. Telomeres and the areas adjacent to centromeres are composed of heterochromatin.

9. (a) Since there are 200 base pairs per nucleosome (as defined in this problem) and 10^9 base pairs, there would be 5×10^6 nucleosomes. **(b)** Given that there are six nucleosomes per solenoid, there would be 7.83×10^5 solenoids. **(c)** Since there are 5×10^6 nucleosomes and nine histones (including H1) per nucleosome, there must be $9(5 \times 10^6)$ histone molecules: 4.5×10^7. **(d)** Since there are 10^9 base pairs present and each base pair is 3.4 Å the overall length of the DNA is 3.4×10^9 Å. Dividing this value by the packing ratio (50) gives 6.8×10^7 Å .

10. The first step of this solution is to convert all of the given values to cubic Å remembering that 1 mm = 10,000 Å. Using the formula πr^2 for the area of a circle and $4/3 \pi r^3$ for the volume of a sphere, the following calculations apply:

Volume of DNA: 3.14×10 Å $\times 10$ Å $\times$ $(50 \times 10^4$ Å$) = 1.57 \times 10^8$ Å^3

Volume of capsid: $4/3 (3.14 \times 400$ Å $\times 400$ Å $\times$ 400 Å$) = 2.67 \times 10^8$ Å^3

Because the capsid head has a greater volume than the volume of DNA, the DNA will fit into the capsid.

11. One base pair occupies 0.34nm, therefore the equation would be as follows:

$52\mu m/(0.34nm/bp) \times 1000nm/mm =$

$\qquad$ 152,941 base pairs

12. Volume of the nucleus = $4/3 \pi r^3$

$\qquad = 4/3 \times 3.14 \times (5 \times 10^3 nm)^3$

$\qquad = 5.23 \times 10^{11} nm^3$

Volume of the chromosome = $\pi r^2 \times$ length

$\qquad = 3.14 \times 5.5nm \times 5.5nm \times (2 \times 10^9 nm)$

$\qquad = 1.9 \times 10^{11} nm^3$

Therefore, the percentage of the volume of the nucleus occupied by the chromatin is

$\qquad = 1.9 \times 10^{11} nm^3 / 5.23 \times 10^{11} nm^3 \times 100$

$\qquad =$ about 36.3%

13. The basic issue here is to determine whether all the loci which are not from the mother can come from the father. That is, does the father's genotype provide the child's loci that are not provided by the mother. For example, at the *D9S302* locus, the mother contributed the child's 31 locus and the alleged father could have contributed the child's 32 locus. Since other men also have the 32 locus, one can't say that the alleged father is the father, rather, one could say that the alleged father cannot be excluded as the source of the sperm that produced the child.

Chapter 5: The Genetic Code and Transcription

Concept Areas	Corresponding Problems
Genetic Code	1, 5, 9, 10, 12, 13, 15, 17
Deciphering the Code	3, 4, 6, 7, 8, 11
Characteristics of the Code	2, 14, 16, 26
Information Flow	18, 19, 20, 21, 22, 23, 24, 25, 27
RNA Structure	14, 16

Vocabulary: Organization and Listing of Terms and Concepts

Structures and Substances

Codon

  triplet

Messenger RNA

  *r*II, β cistron

Proflavin

Phage T4

Strain B of *E. coli*

Strain K12 of *E. coli*

Tobacco mosaic virus (TMV)

  adaptor molecule

Polynucleotide phosphorylase

  random assembly of nucleotides

Homopolymer codes

  RNA homopolymers

  RNA heteropolymers

  anticodon

N-formylmethionine (fmet)

Ribosome

RNA polymerase

  holoenzyme

  (α, β, β', σ)

  consensus sequences

  Pribnow box (-10) TATA

  -35 region

  termination factor (rho)

Messenger RNA (mRNA)

  polycistronic mRNA

  monocistronic mRNA

RNA polymerase (eukaryotic) - I, II, III

  heterogeneous nuclear RNA (hnRNA)

heterogeneous nuclear
ribonucleoprotein (hnRNP)

α-amanitin

nucleoside triphosphates (NTPs)

nucleoside monophosphates (NMPs)

nucleotides

promoters (promoter sequences)

Consensus sequences

adenine and thymine richness

cis-acting elements

Goldberg-Hogness (-30, TATA box)

CCAAT sequence

enhancers

trans-acting factors

transcription factors

TATA-factor (TFIIA, B)

TATA-binding protein (TBP)

pre-mRNAs

split genes (intervening sequences)

introns

exons

heteroduplexes

β-globin gene

ovalbumin gene

dystrophin

poly-A

cap (7mG)

5' to 5'

isoforms

double-stranded RNA adenosine
deaminase

ADAR (adenosine deaminase acting on
RNA)

Processes/Methods

Transcription, Translation

Frameshifts

Cell-free protein-synthesizing system

ribosomes, tRNAs, amino acids, *etc.*

artificial mRNAs

Triplet binding assay

Suppression

Overlapping genes

multiple initiation

Transcription

RNA polymerase II

cleft, clamp

active center

abortive transcription

lid, pore

(mRNA, evidence for)

Chapter 5 The Genetic Code and Transcription

template binding

    template strand, partner strand

denaturation (unwinding)

DNA footprinting

initiation

chain elongation (5' to 3')

chain termination

gene amplification

RNA processing

  split genes

  post-transcriptional changes

    poly-A (3'), cap (5')

  mechanisms

    rRNA self-excision (ribozyme)

    spliceosome

      snRNAs, snurps (snRNP)

    branch point

    alternative splicing

      isoform

RNA editing

  substitution

  insertion/deletion

  guide RNA (gRNA)

Concepts

Genetic code

  triplet codon

    frameshift mutations (r_{II})

    (+++)(- - -)

      nonsense triplets

    codon assignments

      artificial mRNAs

      triplet binding assay

      repeating copolymers

    ordered codons

    confirmation of codon assignments

      MS2 sequencing, colinearity

  unambiguous

  degenerate, wobble

    support for degenerate code

  punctuation

    start, AUG, GUG (rare)

    stop, UAA, UAG, UGA

  nonoverlapping

    support for nonoverlapping code

  universal, exceptions

  ordered

Hypotheses

    adaptor molecule

    messenger RNA

    wobble hypothesis

        pattern of degeneracy

Information flow (F5.1)

    transcription

        primary transcript

        an intermediate molecule

RNA polymerases (I, II, III)

    translation (F5.2)

gene amplification

RNA splicing

    beta-globin gene

    ovalbumin gene

    pro-α-2(I) collagen

    mechanisms (4 classes)

Alternative splicing, RNA editing

Comparisons (eukaryotic, prokaryotic)

F5.1 Illustration of the processes, transcription and translation, involved in protein synthesis. Such relationships are often called the Central Dogma.

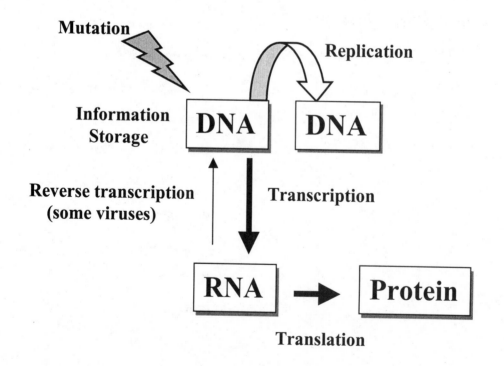

F5.2 Illustration of transcription in prokaryotes coupled with translation. Transcription involves production of RNA from a DNA template.

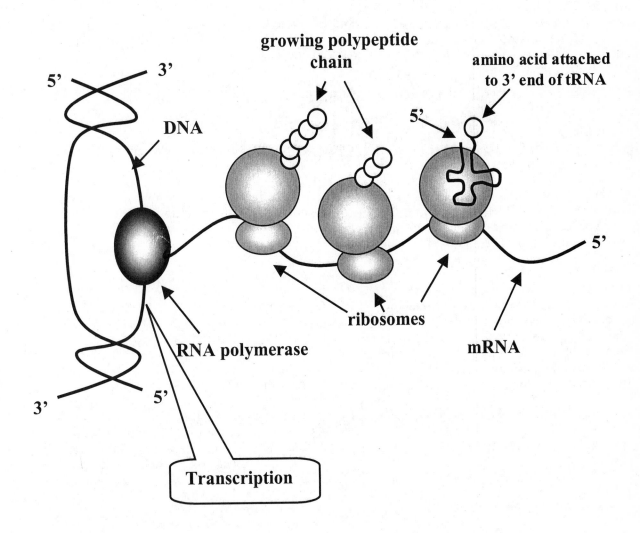

Solutions to Problems and Discussion Questions

1. In eukaryotes, protein synthesis occurs primarily in the cytoplasm, far from the location of DNA and the encoded information. In addition, while some of the basic amino acids would be able to associate directly with DNA, the acidic amino acids would be unable to do so. Thus some sort of "adaptor" system was needed for DNA to direct amino acid assembly.

2. No. The the term "reading frame" refers to the number of bases contained in each codon. The reason that (+++) or (- - -) restored the reading frame is because the code is triplet. By having the (+++) or (- - -), the translation system is "out of phase" until the third "+" or "-" is encountered. If the code contained six nucleotides (a sextuplet code), then the translation system is "out of phase" until the sixth "+" or "-" is encountered.

3. (a) The way to determine the fraction which each triplet will occur with a random incorporation system is to determine the likelihood that each base will occur in each position of the codon (first, second, third), then multiply the individual probabilities (fractions) for a final probability (fraction).

GGG	= 3/4 X 3/4 X 3/4 =	27/64
GGC	= 3/4 X 3/4 X 1/4 =	9/64
GCG	= 3/4 X 1/4 X 3/4 =	9/64
CGG	= 1/4 X 3/4 X 3/4 =	9/64
CCG	= 1/4 X 1/4 X 3/4 =	3/64
CGC	= 1/4 X 3/4 X 1/4 =	3/64
GCC	= 3/4 X 1/4 X 1/4 =	3/64
CCC	= 1/4 X 1/4 X 1/4 =	1/64

(b) Glycine:

GGG and one G_2C (adds up to 36/64)

Alanine:

one G_2C and one C_2G (adds up to 12/64)

Arginine:

one G_2C and one C_2G (adds up to 12/64)

Proline:

one C_2G and CCC (adds up to 4/64)

(c) With the wobble hypothesis, variation can occur in the third position of each codon.

Glycine: GGG, GGC

Alanine: CGG, GCC, CGC, GCG

Arginine: GCG, GCC, CGC, CGG

Proline: CCC,CCG

4. Assume that you have introduced a copolymer (ACACACAC. . .) to a cell free protein synthesizing system. There are two possibilities for establishing the reading frames: ACA, if one starts at the first base and CAC if one starts at the second base. These would code for two different amino acids (ACA = threonine; CAC = histidine) and would produce repeating polypeptides which would alternate *thr-his-thr-his*. . . or *his-thr-his-thr*. . .

Because of a triplet code, a trinucleotide sequence will, once initiated, remain in the same reading frame and produce the same code all along the sequence regardless of the initiation site.

Chapter 5 The Genetic Code and Transcription

Given the sequence CUACUACUACUA, notice the different reading frames producing three different sequences each containing the same amino acid.

Codons: CUA   CUA   CUA   CUA. . .

Amino Acids: leu   leu   leu   leu. . .

              UAC   UAC   UAC   UAC. . .
              tyr   tyr   tyr   tyr. . .

              ACU   ACU   ACU   ACU. . .
              thr   thr   thr   thr. . .

If a tetranucleotide is used, such as ACGUACGUACGU...

Codons: ACG   UAC   GUA   CGU   ACG

Amino Acids:thr   tyr   val   arg   thr

              CGU   ACG   UAC   GUA   CGU

              arg   thr   tyr   val   arg

              GUA   CGU   ACG   UAC   GUA

              val   arg   thr   tyr   val

              UAC   GUA   CGU   ACG   UAC

              tyr   val   arg   thr   tyr

Notice that the sequences are the same except that the starting amino acid changes.

5. The UUACUUACUUAC tetranucleotide sequence will produce the following triplets depending on the initiation point: UUA = leu; UAC = tyr; ACU = thr; CUU = leu. Notice that because of the degenerate code, two codons correspond to the amino acid leucine.

The UAUCUAUCUAUC tetranucleotide sequence will produce the following triplets depending on the initiation point: UAU = tyr; AUC = ileu; UCU = ser; CUA = leu. Notice that in this case, degeneracy is not revealed and all the codons produce unique amino acids.

6. From the repeating polymer ACACA. . . one can say that threonine is either CAC or ACA. From the polymer CAACAA. . . with ACACA. . . , ACA is the only codon in common. Therefore, threonine would have the codon ACA.

7. As in the previous problem the procedure is to find those sequences which are the same for the first two bases but which vary in the third base. Given that AGG = arg, then information from the AG copolymer indicates that AGA also codes for arg and GAG must therefore code for glu.

Coupling this information with that of the AAG copolymer, GAA must also code for glu, and AAG must code for lys.

8. The basis of the technique is that if a trinucleotide contains bases (a codon) which are complementary to the anticodon of a charged tRNA, a relatively large complex is formed which contains the ribosome, the tRNA, and the trinucleotide. This complex is trapped in the filter whereas the components by themselves are not trapped. If the amino acid on a charged, trapped tRNA is radioactive, then the filter becomes radioactive.

9. List the substitutions, then from the code table, apply the codons to the original amino acids. Select codons which provide single base changes.

Original		Substitutions
threonine	----->	*alanine*
AC(U,C,A, or G)		GC(U,C,A, or G)
glycine	----->	*serine*
GG(U or C)		AG(U or C)
isoleucine	----->	*valine*
AU(U, C or A)		GU(U, C or A)

10. Apply the most conservative pathway of change.

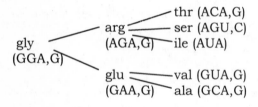

11. The enzyme generally functions in the degradation of RNA, however in an *in vitro* environment, with high concentrations of the ribonucleoside diphosphates, the direction of the reaction can be forced toward polymerization.

   In vivo, the concentration of ribonucleoside diphosphates is low and the degradative process is favored.

12. Because Poly U is complementary to Poly A, double-stranded structures will be formed. In order for an RNA to serve as a messenger RNA it must be single-stranded thereby exposing the bases for interaction with ribosomal subunits and tRNAs.

13. Applying the coding dictionary, the following sequences are "decoded"

   Sequence 1: met-pro-asp-tyr-ser-(term)

   Sequence 2: met-pro-asp-(term)

The 12th base (a uracil) is deleted from Sequence #1 thereby causing a frameshift mutation.

14. (a)

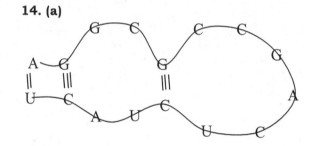

(b) TCCGCGGCTGAGATGA (use complementary bases, substituting T for U)

(c) GCU

(d) Assuming that the AGG. . . is the 5' end of the mRNA, then the sequence would be

   arg-arg-arg-leu-tyr

15. Given the sequence GGA, by changing each of the bases to the remaining three bases, then checking the code table, one can determine whether amino acid substitutions will occur.

G G A gly	G G U gly
U G A **term**	G G C gly
C G A **arg**	G G A gly
A G A **arg**	G G G gly
G U A **val**	U G U **cys**
G C A **ala**	C G U **arg**
G A A **glu**	A G U **ser**
G G U gly	G U U **val**
G G C gly	G C U **ala**
G G A gly	G A U **asp**

48

Chapter 5 The Genetic Code and Transcription

16. (a) Starting from the 5' end and locating the AUG triplets one finds two initiation sites leading to the following two sequences:

met-his-tyr-glu-thr-leu-gly

met-arg-pro-leu-gly

(b) In the shorter of the two reading sequences (the one using the internal AUG triplet), a UGA triplet was introduced at the second codon. While not in the reading frames of the longer polypeptide (using the first AUG codon) the UGA triplet eliminates the product starting at the second initiation codon.

17. By examining the coding dictionary one will notice that the number of codons for each particular amino acid (synonyms) is directly related to the frequency of amino acid incorporation stated in the problem.

18. The central dogma of molecular genetics and to some extent, all of biology, states that DNA produces, through transcription, RNA, which is "decoded" (during translation) to produce proteins. See F5.2 for a graphic description.

19. Several observations indicated that a "messenger" molecule exists. First, DNA, the genetic material, is located in the nucleus of a eukaryotic cell whereas protein synthesis occurs in the cytoplasm. DNA, therefore, does not directly participate in protein synthesis. Second, RNA, which is chemically similar to DNA is synthesized in the nucleus of eukaryotic cells. Much of the RNA migrates to the cytoplasm, the site of protein synthesis. Third, there is generally a direct correlation between the amounts of RNA and protein in a cell. More direct support was derived from experiments showing that an RNA other than that found in ribosomes was involved in protein synthesis and shortly after phage infection, an RNA species is produced which is complementary to phage DNA.

20. RNA polymerase from *E. coli* is a complex, large (almost 500,000 daltons) molecule composed of subunits ($\alpha,\beta,\beta',\sigma$) in the proportion $\alpha2,\beta,\beta'$, σ for the holoenzyme. The β subunit provides catalytic function while the sigma (σ) subunit is involved in recognition of specific promoters. The core enzyme is the protein without the sigma.

21. Ribonucleoside triphosphates and a DNA template in the presence of RNA polymerase and a divalent cation (Mg^{++}) produce a ribonucleoside monophosphate polymer, DNA, and pyrophospate (diphosphate). Equimolar amounts of precursor ribonucleoside triphosphates, and product ribonucleoside monophosphates and pyrophosphates (disphosphates) are formed. In *E. coli* transcription and translation can occur simultaneously. Ribosomes add to the 5' end nascent mRNA and progress to the 3' end during translation. While transcription/translation can be "visualized" in the *E. coli* (F5.2), the predominant components "visualized" are the strings of ribosomes (polysomes).

22. While some folding (from complementary base pairing) may occur with mRNA molecules, they generally exist as single-stranded structures which are quite labile. Eukaryotic mRNAs are generally processed such that the 5' end is "capped" and the 3' end has a considerable string of adenine bases. It is thought that these features protect the mRNAs from degradation. Such stability of eukaryotic mRNAs probably evolved with the differentiation of nuclear and cytoplasmic functions. Because prokaryotic cells exist in a more unstable environment (nutritionally and physically, for example) than many cells of multicellular organisms, rapid genetic response to environmental change is likely to be adaptive. To accomplish such rapid responses, a labile gene product (mRNA) is advantageous. A pancreatic cell, which is developmentally stable (differentiated) and existing in a relatively stable environment could produce more insulin on stable mRNAs for a given transcriptional rate.

49

Chapter 5 The Genetic Code and Transcription

23. Apply complementary bases, substituting U for T:

(a)

Sequence 1: GAAAAAACGGUA

Sequence 2: UGUAGUUAUUGA

Sequence 3: AUGUUCCCAAGA

(b)

Sequence 1: *glu-lys-thr-val*

Sequence 2: *cys-ser-tyr*

Sequence 3: *met-phe-pro-arg*

(c) Apply complementary bases:

GAAAAAACGGTA

24. First, compute the frequency (percentages would be easiest to compare) for each of the random codons.

For 4/5 C: 1/5 A:

CCC= 4/5 X 4/5 X 4/5 = 64/125 (51.2%)

C_2A = 3(4/5 X 4/5 X 1/5) = 48/125 (38.4%)

CA_2 = 3(4/5 X 1/5 X 1/5) = 12/125 (9.6%)

AAA = 1/5 X 1/5 X 1/5 = 1/125 (0.8%)

For 4/5 A: 1/5 C:

AAA = 4/5 X 4/5 X 4/5 = 64/125 (51.2%)

A_2C = 3(4/5 X 4/5 X 1/5) = 48/125 (38.4%)

AC_2 = 3(4/5 X 1/5 X 1/5) = 12/125 (9.6%)

CCC = 1/5 X 1/5 X 1/5 = 1/125 (0.8%)

Proline: C_3, and one of the C_2A triplets

Histidine: one of the C_2A triplets

Threonine: one C_2A triplet,

and one A_2C triplet

Glutamine: one of the A_2C triplets

Asparagine: one of the A_2C triplets

Lysine: A_3

25.

(a) #1: *nonsense mutation*
#2: *missense mutation*
#3: *frameshift mutation*

(b) #1: mutation in third position to A or G
#2: change U to C in third triplet
#3: removal of a G in the UGG triplet (trp)

(c) termination

(d) All of the amino acids can be assigned specific triplets including the third base of each triplet. Compare the sequences for the wild type and mutant#2. After removal of a G in the UUG triplet of tryptophan, the frameshift mutation shifts the first base of the following triplet to the third (often ambiguous) base of the previous triplet. The only tricky solution is with serine which has six triplet possibilities, but it can still be resolved.

AUG UGG UAU CGU GGU AGU CCA ACA

(e) The mutation may be in a promoter or enhancer, although many posttranscriptional alterations are possible. Depending on the gene and the organism, the mutation may be in an intron/exon splice site, *etc.*

26. (a,b) Use the code table to determine the number of triplets which code each amino acid, then construct a graph and plot such as this one below:

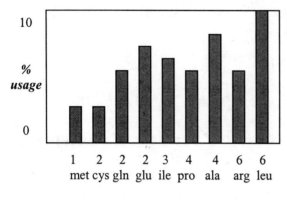

codons

(c) There appears to be a weak correlation between the relative frequency of amino acid usage and the number of triplets for each.

(d) To continue to investigate this issue one might examine additional amino acids in a similar manner.   In addition, different phylogenetic groups use code synonyms differently. It may be possible to find situations in which the relationship are more extreme. One might also examine more proteins to determine whether such a weak correlation is stronger with different proteins.

27. Consider the following diagram below representing the possibility described in the problem. If the promoter was not transcribed, as is the case of typical protein-coding genes, retrotransposition would produce a "daughter" *Alu* void of a promoter. Such an *Alu* would be a "dead-end" because it would not be capable of giving rise to its own *Alu* sequences. Perhaps an *Alu* sequence inserted 3' to a promoter would produce daughters, but this would likely be rare.

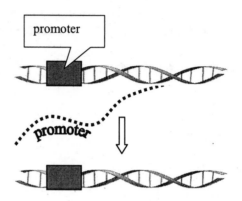

Chapter 6: Translation and Proteins

Concept Areas	Corresponding Problems
Translation	1, 4, 21, 22, 26, 28
RNAs	2, 3, 6, 7
Information Flow	5, 27
One-gene: One-enzyme	8, 9, 10, 14
Pathways	12, 13
Proteins	11, 16, 17, 18, 19, 20, 23, 24, 25, 28

Vocabulary: Organization and Listing of Terms and Concepts

Structures and Substances

Polypeptide, protein

Signal sequence

Chaperone

Ribosome

   monosome

rRNA

rDNA

   5S, 16S, 23S RNA (single transcript)

      31 proteins

   5.8S, 18S, 28S RNA (single transcript)

      5S

      50 proteins

   tandem repeats and spacer DNA

   clustered on chromosomes:
      13, 14, 15, 21, 22

   subunits

ribosomal proteins

   31, 21 (prokaryotic)

   50, 33 (eukaryotic)

Ribosome complex

   peptidyl (P site)

   aminoacyl (A site)

   exit (E site)

   tunnel

   peptidyl transferase

Transfer RNAs - tRNA

   75-90 nucleotides

   unusual bases

   cloverleaf model

   cognate amino acid

      anticodon, codon

      ...pCpCpA (3')

      ...pG (5')

Chapter 6 Translation and Proteins

aminoacyl tRNA synthetases

    charging

    activated form

        (aminoacyladenylic acid)

    isoaccepting tRNAs

Initiation factors

    initiation complex

    Shine-Delgarno sequence

Formylmethionine (N-formylmethionine)

    tRNA^{fmet} (unique tRNA$_i$^{met})

Elongation factors

GTP-dependent release factors

    termination (nonsense)

    stop codons UAA, UAG, UGA

Polyribosomes (polysomes)

mRNA

    Heterogeneous RNA (hnRNA)

        poly-A

        cap (7mG)

           5' to 5'

        Kozak sequences

        5'-ACCAUGG

        Sec61

Inborn Errors of Metabolism

    alkaptonuria

        homogentisic acid

        (alkapton 2,5-dihydrophenylacetic acid)

    phenylketonuria

        phenylalanine hydroxylase

    Neurospora

    arginine

    citrulline

    ornithine

    sickle-cell anemia (trait)

        HbA, HbS, HbA$_2$

        heme group

        globin portion

        alpha α

        beta β

        delta δ

        Gower 1

           zeta ζ

           epsilon ε

        HbF

           gamma γ

        tryptophan synthetase

Amino acid

   carboxyl group

   amino group

   R (radical) group

   central carbon

   hydrophobic

   polar (hydrophilic)

      negative, positive

Peptide bond

   dipeptide, tripeptide

Primary structure

Secondary structure

   α-helix

   β-pleated sheet

   fibroin

Tertiary structure

   myoglobin

Quaternary structure

   collagen

   keratin

   actin, myosin

   immunoglobin

   transport proteins

      hemoglobin, myoglobin

   enzyme (active site)

hormone, receptor

anabolic, catabolic

   energy of activation

protein domain

LDL receptor protein

epidermal growth factor

Processes/Methods

Transcription, Translation

RNA processing

   post-transcriptional modification

Translation

   codon, anticodon

   tRNA charging

   aminoacyl tRNA synthetases

   chain initiation

   chain elongation

   translocation

   chain termination, UAG, UAA, UGA

Simultaneous transcription and translation (Prokaryotes, F6.1)

X-ray diffraction

RNA-protein bridge

wobble hypothesis

Starch gel electrophoresis

    Colinear relationships

    Exon shuffling

    Posttranslational modification

        protein targeting

        modification, trimming

        complex formation

Concepts

Information flow

Transcription, translation (F6.1)

Comparisons (eukaryotic, prokaryotic)

One-gene: One-enzyme hypothesis

One-gene: One-protein

One-gene: One-polypeptide chain

Pathway analysis

Colinearity, Protein Structure

Posttranslational modification

Structure/function relationships

F6.1 Polarity constraints associated with simultaneous transcription and translation in prokaryotes. The RNA polymerase is moving downward (arrow) in this sketch.

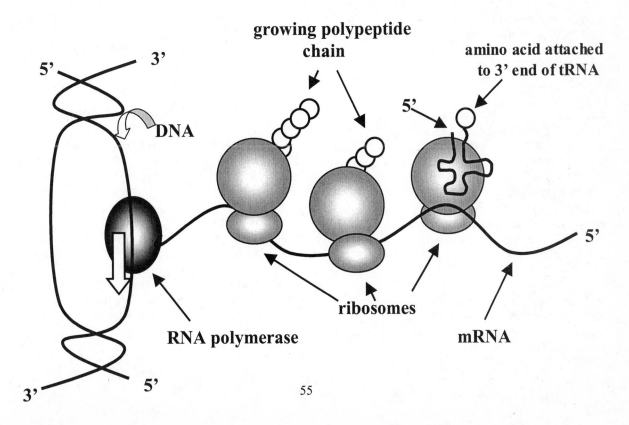

Solutions to Problems and Discussion Questions

1. A functional polyribosome will contain the following components: mRNA, charged tRNA, large and small ribosomal subunits, elongation and perhaps initiation factors, peptidyl transferase, GTP, Mg^{++}, nascent proteins, possibly GTP-dependent release factors.

2. Transfer RNAs are "adaptor" molecules in that they provide a way for amino acids to interact with sequences of bases in nucleic acids. Amino acids are specifically and individually attached to the 3' end of tRNAs which possess a three-base sequence (the anticodon) to base-pair with three bases of mRNA. Messenger RNA, on the other hand, contains a copy of the triplet codes which are stored in DNA. The sequences of bases in mRNA interact, three at a time, with the anticodons of tRNAs.

Enzymes involved in transcription include the following: RNA polymerase (*E. coli*), and RNA polymerase I, II, III (eukaryotes). Those involved in translation include the following: aminoacyl tRNA synthetases, peptidyl transferase, and GTP-dependent release factors.

3. It was reasoned that there would not be sufficient affinity between amino acids and nucleic acids to account for protein synthesis. For example, acidic amino acids would not be attracted to nucleic acids. With an adaptor molecule, specific hydrogen bonding could occur between nucleic acids, and specific covalent bonding could occur between an amino acid and a nucleic acid tRNA.

4. The sequence of base triplets in mRNA constitutes the sequence of codons. A three-base portion of the tRNA constitutes the anticodon.

5. Since there are three nucleotides which code for each amino acid, there would be 423 code letters (nucleotides), 426 including a termination codon. This assumes that other features, such as the polyA tail, the 5'cap, and non-coding leader sequences are omitted. Dividing 20 by 0.34, gives the number of nucleotides (about 59) occupied by a ribosome. Dividing 59 by three gives the approximate number of triplet codes: approximately 20.

6. The steps involved in tRNA charging are outlined in the K/C text. An amino acid in the presence of ATP, Mg^{++}, and a specific aminoacyl synthetase produces an amino acid-AMP enzyme complex ($+ PP_i$). This complex interacts with a specific tRNA to produce the aminoacyl tRNA.

7. The four sites in tRNA which provide for specific recognition are the following: attachment of the specific amino acid, interaction with the aminoacyl tRNA synthetase, interaction with the ribosome, and interaction with the codon (anticodon).

8. Phenylalanine is an amino acid which, like other amino acids, is required for protein synthesis. While too much phenylalanine and its derivatives cause PKU in phenylketonurics, too little will restrict protein synthesis.

9. Both phenylalanine and tyrosine can be obtained from the diet. Even though individuals with PKU can not convert phenylalanine to tyrosine, it is obtained from the diet.

10. Tyrosine is a precursor to melanin, skin pigment. Individuals with PKU fail to convert phenylalanine to tyrosine and even though tyrosine is obtained from the diet, at the population level, individuals with PKU have a tendency for less skin pigmentation.

11. When an expectant mother returns to consumption of phenylalanine in her diet, she subjects her baby to higher than normal levels of phenylalanine throughout its development. Since increased phenylalanine is toxic many (approximately 90%) newborns are severely and irreversibly retarded at birth. Expectant mothers (who are genetically phenylketonurics) should return to a low phenylalanine intake during pregnancy.

12. The best way to approach these types of problems, especially when the data are organized in the form given, is to realize that the substance (supplement) which "repairs" a strain, as indicated by a (+), is *after* the metabolic block for that strain. In addition, and most important, is that the substance which "repairs" the highest number of strains is either *the end product* or is *closest to the end product*. Looking at the table, notice that the supplement tryptophan "repairs" all the strains. Therefore it must be at the end of the pathway or at least after all the metabolic blocks (defined by each mutation). Indole "repairs" the next highest number of strains (3) therefore it must be second from the end. Indole glycerol phosphate "repairs" two of the four strains so it is third from the end. Anthranilic acid "repairs" the least number of strains, so it must be early (first) in the pathway.

Minimal medium is void of supplements and mutant strains involving this pathway would not be expected to grow (or be "repaired'). The pathway therefore would be as follows:

$$\boxed{AA\text{-----}>IGP\text{-----}>I\text{-----}>TRY}$$

To assign the various mutations to the pathway, keep in mind that if a supplement "repairs" a given mutant, the supplement must be after the metabolic block. Applying this rationale to the above pathway, the metabolic blocks are created at the following locations.

$$\boxed{\begin{array}{cccc} trp\text{-}8 & trp\text{-}2 & trp\text{-}3 & trp\text{-}1 \\ \text{precursor-} & \rightarrow AA\text{-} & \rightarrow IGP & \text{--->I--} & \rightarrow TRY \end{array}}$$

13. In general the rationale for working with a branched chain pathway is similar to that stated in the previous problem. Since thiamine "repairs" each of the mutant strains, it must, as stated in the problem, be the final synthetic product.

Remembering the "one-gene:one-enzyme" statement, each metabolic block should only occur in one place, so even though pyrimidine and thiazole supplements each "repair" only one strain, they will not occupy the same step; rather a branched pathway is suggested. Consider that pyrimidine and thiazole are products of distinct pathways and that both are needed to produce the end product, thiamine, as indicated below:

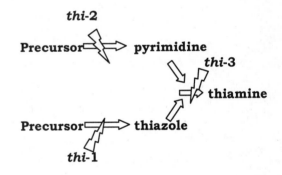

14. The fact that enzymes are a subclass of the general term *protein*, a *one-gene:one-protein* statement might seem to be more appropriate. However, some proteins are made up of subunits, each different type of subunit (polypeptide chain) being under the control of a different gene. Under this circumstance, the *one-gene:one-polypeptide* might be more reasonable.

It turns out that many functions of cells and organisms are controlled by stretches of DNA which either produce no protein product (operator and promoter regions, for example) or have more than one function as in the case of overlapping genes and differential mRNA splicing. A simple statement regarding the relationship of a stretch of DNA to its physical product is difficult to formulate.

15. The electrophoretic mobility of a protein is based on a variety of factors, primarily the net charge of the protein and to some extent, the conformation in the electrophoretic environment. Both are based on the type and sequence (primary structure) of the component amino acids of a protein.

The interactions (hydrogen bonds) of the components of the peptide bonds, hydrophobic, hydrophilic, and covalent interactions (as well as others) are all dependent on the original sequence of amino acids and take part in determining the final conformation of a protein. A change in the electrophoretic mobility of a protein would therefore indicate that the amino acid sequence had been changed.

16. The following types of normal hemoglobin are presented in the text:

Hemoglobin	Polypeptide chains
HbA	2α2β (alpha, beta)
HbA₂	2α2δ (alpha, delta)
HbF	2α2γ (alpha, gamma)
Gower 1	2ζ2ε (zeta, epsilon)

The *alpha* and *beta* chains contain 141 and 146 amino acids, respectively. The *zeta* chain is similar to the alpha chain while the other chains are like the beta chain.

17. Sickle-cell anemia is coined a *molecular* disease because it is well understood at the molecular level; at the level of a base change in DNA which leads to an amino acid change in the β chain of hemoglobin. It is a *genetic* disease in that it is inherited from one generation to the next. It is not contagious as might be the case of a disease caused by a microorganism. Diseases caused by microorganisms may not necessarily follow family blood lines whereas genetic diseases do.

18. In the late 1940's Pauling demonstrated a difference in the electrophoretic mobility of HbA and HbS (sickle-cell hemoglobin) and concluded that the difference had a chemical basis. Ingram determined that the chemical change occurs in the primary structure of the globin portion of the molecule using the fingerprinting technique. He found a change in the 6th amino acid in the β chain.

19. It is possible for an amino acid to change without changing the electrophoretic mobility of a protein under standard conditions. If the amino acid is substituted with an amino acid of like charge and similar structure there is a chance that factors which influence electrophoretic mobility (primarily net charge) will not be altered. Other techniques such as chromatography of digested peptides may detect subtle amino acid differences.

20. *Colinearity* refers to the sequential arrangement of subunits, amino acids and nitrogenous bases, in proteins and DNA, respectively. Sequencing of genes and products in MS2 phage and studies on mutations in the A subunit of the *tryptophan synthetase* gene indicate a colinear relationship.

21. "Fine-mapping" meaning precise mapping of mutations *within* a gene, is possible in some phage systems because many recombinants can often be generated relatively easily. Having the precise intragenic location of mutations as well as the ability to isolate the products, especially mutant products, allows scientists to compare the locations of lesions within genes.

Mutations occurring early in a gene will produce proteins with defects near the N-terminus. In this problem, the lesions cause chain termination, therefore the nearer the mutations are to the 5' end of the mRNA, the shorter will be the polypeptide product.

22. As stated in the text, the four levels of protein structure are the following:

Primary: the linear arrangement or sequence of amino acids. This sequence determines the higher level structures.

Secondary: α-helix and β-pleated sheet structures generated by hydrogen bonds between components of the peptide bond.

Tertiary: folding which occurs as a result of interactions of the amino acid side chains. These interactions include, but are not limited to the following: covalent disulfide bonds between cysteine residues; interactions of hydrophilic side chains with water; interactions of hydrophobic side chains with each other.

Quaternary: the association of two (dimer) or more polypeptide chains. Called *oligomeric,* such a protein is made up of more than one protein.

23. There are probably as many different types of proteins as there are different types of structures and functions in living systems. Your text lists the following:

Oxygen transport: hemoglobin, myoglobin
Structural: collagen, keratin, histones
Contractile: actin, myosin
Immune system: immunoglobins
Cross-membrane transport: a variety of proteins in and around membranes, such as receptor proteins.
Regulatory: hormones, perhaps histones
Catalytic: enzymes

24. Enzymes function to regulate catabolic and anabolic activities of cells. They influence (lower) the *energy of activation* thus allowing chemical reactions to occur under conditions which are compatible with living systems. Enzymes possess active sites and/or other domains which are sensitive to the environment. The active site is considered to be a crevice, or pit, which binds reactants, thus enhancing their interaction. The other domains mentioned above may influence the conformation and therefore function of the active site.

25. Yanofsky's work on the *try* A locus in bacteria involved the mapping of mutations and the finding that a relationship exists between the position of the mutation in a gene and the amino acid change in a protein. Work with the MS2 by Fiers showed, by sequencing of the coat protein (129 amino acids) and the gene (387 nucleotides), a linear relationship as predicted by the code word dictionary.

26. All of the substitutions involve one base change.

27. One can conclude that the amino acid is not involved in recognition of the codon.

28. With the codes for valine being GUU, GUC, GUA, and GUG, single base changes from glutamic acid's GAA and GAG can cause the glu>>>val switch. The normal glutamic acid is a negatively charged amino acid whereas valine carries no net charge and lysine is positively charged. Given these significant charge changes one would predict some, if not considerable, influence on protein structure and function. Such changes could stem from internal changes in folding or interactions with other molecules in the RBC, especially other hemoglobin molecules.

One would expect individuals with HbC to suffer some altered hemoglobin function and perhaps be resistant to malaria as well. In fact, HbC homozygotes suffer mild hemolytic anemia (a benign hemoglobinopathy). The HbC gene is distributed particularly in malarial-infested areas suggesting that some resistance to malaria is conferred.

Recent studies indicate that HbC may be protective against severe forms of malaria, but not to more uncomplicated forms. For more information go to

www.iapac.org/conferences/astmh99/abs8.htm

Chapter 7: Gene Mutation, DNA Repair, and Transposable Elements

Concept Areas	Corresponding Problems
Random and Adaptive Mutations	4, 5
Classes of Mutations	1, 2, 6, 7, 8, 20
Detection of Mutations	3, 21
Spontaneous Mutation Rates	24
Molecular Basis of Mutation	9, 10, 11, 12, 13, 25,
Case Studies	16, 22, 26, 27, 28, 29
Detection of Mutagenicity	17
Repair of DNA	14
UV Radiation and Skin Cancer	22
High-Energy Radiation	12, 13, 15
Site-Directed Mutagenesis	18
Transposable Elements	22, 23, 24

Vocabulary: Organization and Listing of Terms and Concepts

Structures and Substances

Oenothera lamarkiana

lac, Val, salicin

Somatic cells

Gamete forming cells

   germ line

   gametes

   missense

5-Bromouracil

2-Amino purine

Acridine dyes

Acridine orange

Proflavin

Mustard gas

Ethylmethane sulfonate

6-Methylguanine

Apurinic site

ABO antigens

   H substance

   glycosyltransferase

Dystrophin

Pyrimidine dimers

Chapter 7 Gene Mutation, DNA repair, and Transposable Elements

FMR-1

uvr gene product

DNA polymerase I (*pol*A1)

AP endonuclease

DNA glycosylases

XPA, XPF, XPG

TFIIH

*rec*A

*lex*A

Photoreactivation enzyme

Adenine methylase

Heterokaryon

Knockout mice

Transgenic organism

Transposable elements

    insertion sequence

    copia, P

Transposon, transposase

Starch-branching enzyme (SBEI)

Processes/Methods

Chromosomal Aberration

Variation by mutation

    adaptation hypothesis

Luria-Delbruck fluctuation test

adaptive mutations

    lac, Val

    salicin

    spontaneous

replication

repair errors

background radiation

    cosmic sources

    mineral sources

    ultraviolet light

rates

induced

    X-rays

    gametic

        autosomal dominant

        X-linked recessive

        autosomal recessive

    somatic

humans

    molecular basis

        direct sequencing of DNA

ABO antigens

H substance modification

muscular dystrophy

Duchenne muscular dystrophy

Becker muscular dystrophy

dystrophin

trinucleotide

reading frame hypothesis

fragile-X syndrome

repeats

myotonic dystrophy (DM)

chromosome 19

MDPK
(serine-threonine protein kinase)

Huntington disease

spinobulbar muscular dystrophy
(Kennedy disease)

Ames test

Salmonella typhimurium

Molecular basis

base substitution or point mutations

transition

transversion

frameshift

tautomeric shifts (forms)

base analogues

5-bromouracil

2-amino purine

reverse mutation

alkylation

mustard gases

ethylmethane sulfonate

frameshift mutations

acridine dyes (acridines)

acridine orange

proflavin

apurinic sites

deamination

oxygen radicals

H_2O_2

OH^- (hydroxyl radical)

superoxide (O_2^-)

ultraviolet light

pyrimidine dimers

T-T

recA, lexA, uvr

SOS response

high-energy radiation

ionizing radiation

X-rays

gamma radiation

cosmic radiation

free radicals

reactive ions

target theory

Chapter 7 Gene Mutation, DNA repair, and Transposable Elements

intensity of dose

roentgen

Repair

ultraviolet radiation (260nm)

pyrimidine dimers

T-T, C-C, T-C

photoreactivation

photoreactivation enzyme (PRE)

excision repair

base excision repair (BER)

nucleotide excision repair (NER)

uvr gene product

DNA polymerase I

DNA ligase

AP endonuclease

DNA glycosylases

proofreading and mismatch repair

strand discrimination

DNA methylation

GATC sequence

mutH, L, S and *U*

postreplication repair

homologous recombinational repair

(also nonhomologous)

RAD52

double-stranded break repair (DSBR)

xeroderma pigmentosum (XP)

unscheduled DNA synthesis

photoreactivation enzyme

heterokaryon

somatic cell genetics

complementation

Site-directed mutagenesis

Knockout genes

transgenes

Transposable genetic elements

insertion sequence (IS)

terminal repeats

transposon (Tn) elements

bacteriophage *mu*

AC-DS system

transposable controlling elements

open reading frames (ORFs)

transposase

other systems

SBEI gene (plants)

copia (*Drosophila*)

direct terminal repeat (DTR)

inverted terminal repeat (ITR)

P elements (*Drosophila*)

Chapter 7 Gene Mutation, DNA repair, and Transposable Elements

hybrid dysgenesis P, M types

Alu (humans)

SINES, LINES

Concepts

Mutation

basis of organismic diversity (F7.1)

somatic, germ line (F7.2)

dominant autosomal

X-linked, autosomal recessive

morphological

nutritional or biochemical

behavioral

regulatory

lethal

conditional, temperature-sensitive

Sources

exogenous (environmental)

endogenous

spontaneous

Anticipation

Detection of mutations

Molecular diversity of mutations

Repair

Site-directed mutagenesis

Transposable elements

influence on phenotype

structure, origin

Complementation

F7.1 Graphic representation of the relationship of mutation to Darwinian evolutionary theory. Mutation provides the original source of variation on which natural selection operates.

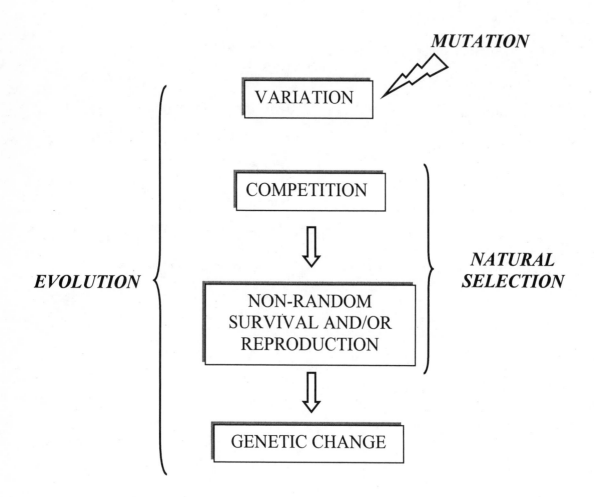

F7.2 Illustration of the difference between somatic and germ line mutations. Somatic mutations are not passed to the next generation whereas those in the germ line may be passed to offspring.

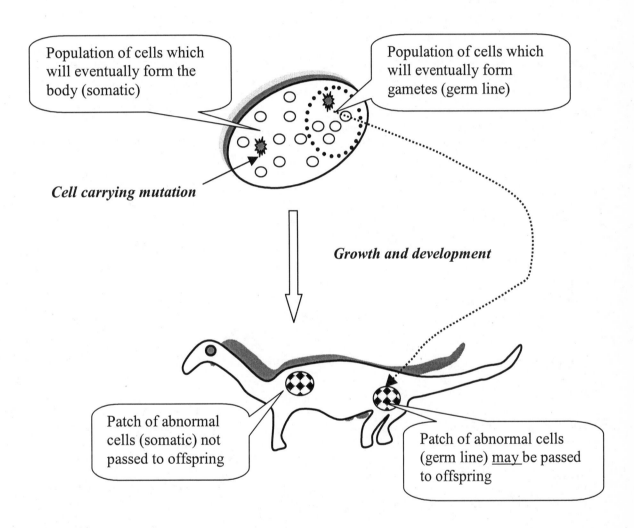

Solutions to Problems and Discussion Questions

1. The term *chromosomal aberration* refers to changes in chromosome number or structure, such as duplications, deletions, inversions, and translocations. A *gene mutation* is a change in the nucleotide sequence of a single gene. Both may include somatic (not passed to offspring) and gametic (potentially passed to offspring) events.

2. Mutations are the "windows" through which geneticists look at the normal function of genes, cells, and organisms. When a mutation occurs it allows the investigator to formulate questions as to the function of the normal allele of that mutation. For example, hemophilia is an inherited blood-clotting disease. Because there are three different inherited forms of the disease, two X-linked and one autosomal, all determined by non-allelic genes, one can say that there are at least three different proteins involved in blood clotting. At a different level, mutations provide "markers" with which biologists can study the genetics and dynamics of populations.

3. Mutagenized conidia can be cultured on complete, then minimal media to isolate strains which are nutritionally deficient (auxotrophs). Individual transfer of such isolates to minimal medium plus various supplements allows one to isolate those mutations which can be "repaired." By crossing an auxotroph to a prototroph one can be certain of the genetic basis of the nutritionally deficient strain. Typical patterns of segregation (4:4, 2:2:2:2, *etc.*) will be observed.

4. It is true that *most* mutations are thought to be deleterious to an organism. A gene is a product of perhaps a billion or so years of evolution and it is only natural to suspect that random changes will probably yield negative results.

However, *all* mutations may not be deleterious. Those few, rare variations which are beneficial will provide a basis for possible differential propagation of the variation. Such changes in gene frequency represent the basis of the evolutionary process. See F7.1 in this book.

5. As stated in the previous problem, a functional sequence of nucleotides, a gene, is likely to be the product of perhaps a billion or so years of evolution. Each gene and its product function in an environment which has also evolved, or co-evolved.

A coordinated output of each gene product is required for life. Deviations from the norm, caused by mutation, are likely to be disruptive because of the complex and interactive environment in which each gene product must function. However, on occasion a beneficial variation occurs.

6. A diploid organism possesses at least two copies of each gene (except for "hemizygous" genes) and in most cases, the amount of product from one gene of each pair is sufficient for production of a normal phenotype. Recall that the condition of "recessive" is defined by the phenotype of the heterozygote. If one unit of output from the normal gene gives the same phenotype as in the normal homozygote, where there are two units of output, the allele is considered "recessive."

Phenotype, if mutant is:

Genotypes	recessive	dominant
wild/wild	wild	wild
wild/mutant	wild	mutant
mutant/mutant	mutant	mutant

7. A *conditional* mutation is one that produces a wild type phenotype under one environmental condition and a mutant phenotype under a different condition. A conditional *lethal* is a gene which under one environmental condition leads to premature death of the organism.

Protein

Abnormal structure and function at 42° C

Normal structure and function at 37° C

8. Mutations constantly occur in both somatic and gametic tissues. When in somatic tissue, the mutation will not be passed to the next generation, however the physiological or structural role of the mutant cell may be compromised, but of little or no impact. If a somatic mutation occurs early in development and if numerous progeny cells are produced from the mutant cell, then a significant tissue mass may have abnormal function. In addition, mutation can alter the normal regulatory aspects of a cell cycle. When such regulation is compromised, then abnormal rates of cell proliferation, cancer, may result. Mutations in the gametic tissue may pose problems for future generations. See F7.2 in this book.

9. Watson and Crick recognized that various tautomeric forms, caused by single proton shifts, could exist for the nitrogenous bases of DNA. Such shifts could result in mutations by allowing hydrogen bonding of normally noncomplementary bases. As stated in the text, important tautomers involve keto-enol pairs for thymine and guanine, and amino-imino pairs for cytosine and adenine.

10. All three of the agents are mutagenic because they cause base substitutions. Deaminating agents oxidatively deaminate bases such that cytosine is converted to uracil and adenine is converted to hypoxanthine. Uracil pairs with adenine and hypoxanthine pairs with cytosine.

Alkylating agents donate an alkyl group to the amino or keto groups of nucleotides, thus altering base-pairing affinities. 6-ethyl guanine acts like adenine, thus pairing with thymine. Base analogues such as 5-bromouracil and 2-amino purine are incorporated as thymine and adenine respectively yet they base pair with guanine and cytosine respectively.

11. Frameshift mutations are likely to change more than one amino acid in a protein product because as the reading frame is shifted, new codons are generated. In addition, there is the possibility that a nonsense triplet could be introduced, thus causing premature chain termination. If a single pyrimidine or purine has been substituted, then only one amino acid is influenced.

12. X-rays are of higher energy and shorter wavelength than UV light. They have greater penetrating ability and can create more disruption of DNA.

13. In contrast to UV light, X-rays penetrate surface layers of cells and thus can affect gamete-forming tissues in multicellular organisms. In addition, X-rays break chromosomes and a variety of chromosomal aberrations can result. Ions and free radicals are formed in the paths of X-rays and these interact with components of DNA to cause mutations. UV light generates pyrimidine dimers, primarily thymine, which distort the normal conformation of DNA and inhibit normal function.

14. *Photoreactivation* can lead to repair of UV-induced damage. An enzyme, photoreactivation enzyme, will absorb a photon of light to cleave thymine dimers. *Excision repair* involves the products of several genes, DNA polymerase I, and DNA ligase to clip out the UV-induced dimer, fill in, and join the phosphodiester backbone in the resulting gap. The excision repair process can be activated by damage which distorts the DNA helix.

Recombinational repair is a system which responds to DNA that has escaped other repair mechanisms at the time of replication. If a gap is created on one of the newly synthesized strands a "rescue operation or SOS response" allows the gap to be filled. Many different gene products are involved in this repair process: *rec*A, *lex*A. In SOS repair, the proofreading by DNA polymerase III is suppressed and this therefore is called an "error-prone system."

15. Because mammography involves the use of X-rays and X-rays are known to be mutagenic, it has been suggested that frequent mammograms may do harm. This subject is presently under considerable debate. At the 2002 World Health Organization conference in Barcelona, Spain, the conclusion was that "mammograms can prevent breast cancer deaths in one in 500 women ages 50 to 69."

16. Each involves a ballooning of tribnucleotide repeats. See the K/C text for a detailed description of the role of trinucleotide repeats in a variety of human diseases. Genetic anticipation is the occurrence of an earlier age of onset of a genetic disease in successive generations.

17. In the *Ames assay* the compound to be tested is incubated with a mammalian liver extract to simulate an *in vivo* environment. This solution is then placed on culture plates with an indicator microorganism, *Salmonella typhimurium*, which is defective in its normal repair processes. The frequency of mutations in the tester strains is an indication of the mutagenicity of the compound.

18. In *site-directed mutagenesis* the goal is usually to change one or more of the nucleotides within a gene. A piece of DNA complementary to the gene being studied is obtained which contains the desired altered base sequence. The DNA with the altered base sequence will hybridize with the original (unaltered) DNA strand and upon replication, two types of duplexes will be formed; one like the original, unaltered sequence, the other containing the altered sequence. The altered gene sequence can be used to transform competent bacteria.

19. *Xeroderma pigmentosum* is a form of human skin cancer caused by perhaps several rare autosomal genes which interfere with the repair of damaged DNA. Studies with heterokaryons provided evidence for complementation, indicating that there may be as many as seven different genes involved. The photoreactivation repair enzyme appears to be involved.

20. Transitions involve pyrimidine --> pyrimidine and purine --> purine substitutions while transversions involve pyrimidine --> purine and purine --> pyrimidine substitutions. The four types of transitions are the following:

A --> G   G --> A
T --> C   C --> T

The eight types of transversions are the following:

A -->   T, C
G -->   T, C
T -->   G, A
C -->   G, A

21. Given that the cells were treated, then allowed to complete one round of replication, the final computation of the mutation rate should be divided by two. The general expression for the mutation rate is the number of mutant cells divided by the total number of cells. In this case the equation would be as follows:

$$\frac{18 \ X \ 10^1}{6 \ X \ 10^7}$$

or $3 \ X \ 10^{-6}$

Now dividing by two (as stated above) gives

$1.5 \ X \ 10^{-6}$.

22. Each organism mentioned in the problem possesses a variety of transposable elements. Bacteria possess insertion sequences (about 800 to 1500 base pairs in length) as well as transposons which are larger. Both are mobile in bacterial, viral, and plasmid DNAs and both have repeated base sequences at their ends. In maize, Barbara McClintock described the genetic behavior of mobile elements (*Ds* and *Ac*). *Ds* can move if *Ac* is present, thus *transposable controlling elements* exist.

An *Ac* element is 4563 base pairs long and similar in structure to some bacterial transposons. Transposons often code for transposase enzymes which are essential for transposition. *Copia* elements in *Drosophila* may be present in numerous copies in the genome and contain direct and inverted terminal repeats. *P* elements, also in *Drosophila* are responsible for a phenomenon called hybrid dysgenesis. Humans possess a variety of transposable elements including the *Alu* family of short interspersed elements (SINES) which are between 200 to 300 base pairs long and may exist in 300,000 copies per genome. Long interspersed elements (LINES) also occur in the human genome and seem to be capable of movement. Such elements share common structural features, are often mobile, and may influence gene activity.

23. It is probable that the IS occupied or interrupted normal function of a controlling region related to the *galactose* genes.

24. It is likely that the reverse transcriptase, in making DNA, provides a DNA segment which is capable of integrating into the yeast chromosome as other types of DNA are known to do.

25. One can determine complementation groupings by placing each heterokaryon giving a "0" into one group and those giving a "+" into a separate group. For instance, *XP1* and *XP2* are placed into the same group because they do not complement each other. However, *XP1* and *XP5* do complement ("+") therefore they are in a different group. Completing such pairings allows one to determine the following groupings:

XP1	XP4	XP5
XP2		XP6
XP3		XP7

The groupings (complementation groups) indicate that there are at least three "genes" which form products necessary for unscheduled DNA synthesis. All of the cell lines which are in the same complementation group are defective in the same product.

26. Your study should include examination of the following short-term aspects: immediate assessment of radiation amounts distributed in a matrix of the bomb sites as well as a control area not receiving bomb-induced radiation, radiation exposure as measured by radiation sickness and evidence of radiation poisoning from tissue samples, abortion rates, birthing rates, and chromosomal studies. Long-term assessment should include: sex-ratio distortion (males being more influenced by X-linked recessive lethals than females), chromosomal studies, birth and abortion rates, cancer frequency and type and, genetic disorders. In each case data should be compared to the control site to see if changes are bomb-related. In addition, to attempt to determine cause-effect, it is often helpful to show a dose response. Thus, by comparing the location of individuals at the time of exposure to the matrix of radiation amounts, one may be able to determine whether those most exposed to radiation suffer the most physiologically and genetically. If a positive correlation is observed, then statistically significant conclusions may be possible.

27. The cystic fibrosis gene produces a complex membrane transport protein which contains several major domains: a highly conserved ATP binding domain, two hydrophobic domains, and a large cytoplasmic domain which probably serves in a regulatory capacity. The protein is like many ATP-dependent transport systems, some of which have been well studied. When a mutation causes clinical symptoms, fluid secretion is decreased and dehydrated mucus accumulates in the lungs and air passages. Mutations which radically alter the structure of the protein (frameshift, splicing, nonsense, deletions, duplications, *etc.*) would probably have more influence on protein function than those which cause relatively minor amino acid substitutions, although this generalization does not always hold true. A protein with multiple functional domains would be expected to react to mutational insult in a variety of ways.

28. (a) For those organisms that generate energy by aerobic respiration, a process occurs which involves the reduction of molecular oxygen. Partially reduced species are produced as intermediates and by-products of such molecular action: O_2^-, H_2O_2, and OH^-. These species are potent electrophilic oxidants that escape mitochondria and attack numerous cellular components. Collectively, these are called reactive oxygen species (ROS).

(b)

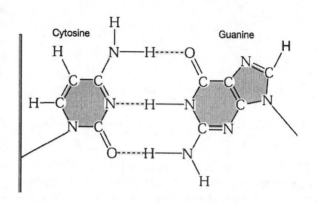

When casually examining the structures in the above diagrams it is not immediately obvious that oxoG:A pairs should occur. However, hydrogen bonding can occur to any other base, including self pairs. Homopurine (A:A, G:G) and heteropurine (A:G) pairs represent anomalous base pairing possibilities even with non-altered bases. While G:C is undoubtedly the most stable, several mispairs are actually stronger than the A-T pair.

Base pairing is complicated by the fact that the purines possess two H-bonding faces, the Watson-Crick face, involving ring positions 1 and 6 for Adenine and, 1, 2, and 6 for Guanine, and the Hoogsteen face involving ring positions 6 and 7. The typical pairing mode is indicated as *wc* where pairing occurs on the Watson-Crick face in the normal orientation, even for the mispair A:G. Alteration of pairing and favoring of the Hoogsteen face can be favored with the alteration generated by oxoGuanine. Indeed, triple helix configurations commonly involve the Hoogsteen face.

(c) If not repaired (see below), in the first round of replication involves the pairing of oxoG to Adenine (see above) while in the next round of replication, Adenine pairs with its normal Thymine. Therefore, if one starts with a G:C pair, one ends up with an A:T pair.

oxoGuanine

(d) It turns out that G:G>T:A transversions are quite commonly found in human cancers and are especially prevalent in the tumor suppressor gene *p53*. Thus, the cellular defense system has been extensively studied. One component is a triphosphatase that cleanses the nucleotide precursor pool by removing the two outermost phosphates from oxo-dGTP. Another involves a DNA glycosylase that initiates repair of misreplicated oxoG:A by hydrolyzing the glycosidic bond linking the adenine base to the sugar. Another is a DNA gycosylase/lyase system that recognizes oxoG opposite cytosine. Of the three systems, the DNA glycosylases are probably the most effective.

29. Since Betazoids have a 4-letter genetic code and the gene is 3,332 nucleotides long, the protein involved must be 833 amino acids in length.

(a) Codon 829 specifies an amino which is very close to the end (carboxyl) of the gene. While a nonsense mutation would terminate translation prematurely, the protein would only be shortened by 5 amino acids. Thus, the protein's ability to fold and perform its cellular function must not be seriously altered. Because of the direction of translation (5' to 3' on the mRNA) the carboxyl terminal amino acids in a protein are the last to be included in folding priorities and are sometimes (often) less significant in determining protein function.

(b) Since the phenotype is mild, this amino acid change does not completely inactivate the protein but it does change its activity to some extent. Perhaps the substitution causes the protein to fold in a slightly aberrant manner, allowing it to have some residual function but preventing it from functioning entirely normally. Additionally, even if the protein folds similar to the wild type protein, charge or structural differences in the protein's active site may be only mildly influenced.

(c) This deletion contains a total of 68 nucleotides, which account for 17 amino acids. Since Betazoids' codons contain four nucleotides, the mRNA reading frames are maintained subsequent to the deletion. Protein function significantly depends on the relative positions of secondary levels of structure: α-helices and β-sheets. If the deleted section is a "benign" linker between more significant protein domains, then perhaps the protein can tolerate the loss of some amino acids in a part of the protein without completely losing its function.

(d) Amino acid specified by codon 192 must be critical to the function of the protein. Altering this amino acid must disrupt a critical region of the protein, thus causing it to lose most or all of its activity. If the protein is an enzyme, this amino acid could be located in its active site and be critical for the ability of the enzyme to bind and/or influence its substrate. One might expect that the amino acid alteration is rather radical such as one sees in the generation of sickle cell anemia. HbS is caused by the substitution of a valine (no net charge) for glutamic acid (negatively charged).

(e) A deletion of 11 base pairs, a number that is not divisible by four, will shift the reading frames subsequent to its location. Even though this deletion is smaller than the deletion discussed above (83-150) and is located in the same region, it causes a reading frame shift and some or all of the amino acids that are added downstream from the mutation may be different from those in the normal protein. The reason that all will not likely change is because of synonyms in the code. There is also the possibility that a nonsense triplet may be introduced in the "out-of-phase" region, thus causing premature chain termination. Because this mutation occurs early in the gene, most of the protein will be affected. This may well explain the severe insensitive phenotype.

Chapter 8: Mitosis and Meiosis

<u>Concept Areas</u>	<u>Corresponding Problems</u>
Cell Structure	1
Homology of Chromosomes	2, 3, 26
Cell Division (Regulation)	8, 23, 24, 25
Mitosis	4, 5, 6, 7, 21, 22
Meiosis	9, 10, 11, 12, 13, 14, 15, 16, 17, 18, 19, 20, 27
Chromosome Structure	21, 22, 26

Vocabulary: Organization and Listing of Terms and Concepts

Structures and Substances

Cells

    plasma membrane

    cell wall

        cellulose, peptidoglycan

    capsule

Diplococcus pneumoniae

        cell coat

        AB, MN antigens

        Duchenne Muscular Dystrophy

           dystrophin

    histocompatibility antigens

    receptor molecules

nucleus, nucleoid

    genetic material (DNA)

    genes

    chromatin

histones

chromosomes

    nucleolus

    nucleolar organizer (NOR)

cytoplasm

    organelles

    cytosol

    endoplasmic reticulum (ER)

    ribosomes

    cytoskeleton

    mitochondria

    chloroplasts

    centrosome

    basal body

    centrioles, spindle fibers

    microtubules, microfilaments

        polar

Chapter 8 Mitosis and Meiosis

tubulin

kinetochore

Chromosomes

chromatin

chromomere

centromere

metacentric

submetacentric

acrocentric

telocentric

p arm, *q* arm

karyotype

locus (loci)

sex-determining chromosomes

X, Y

Mitosis

zygotes

centrosome

kinetochore

spindle fibers

chromatid

sister chromatid

metaphase plate

daughter chromosome

molecular motors

cell plate

middle lamella

cell furrow

Meiosis

synapse

bivalent

synaptonemal complex

Saccharomyces cerevisiae

ZP1

central elements

lateral elements

tetrad

dyad

monad

sister chromatids

nonsister chromatids

chiasma (chiasmata)

spermatogonium

primary spermatocyte

secondary spermatocyte

spermatid

spermatozoa (sperm)

oogonium

primary oocyte

secondary oocyte

first polar body

ootid

second polar body

ova (ovum)

Processes/Methods

Oxidative phases of cell respiration

Photosynthesis

Karyokinesis

Cytokinesis

Mitosis

Cell cycle

  cytokinesis

  interphase

  S phase (replication)

  G1, G2, G0, R

  checkpoints (G1/S, G2/M, M)

   cdc mutations

  cdc kinase

  cdk protein

  cyclin-dependent kinase

  cyclin

  cdk-cyclin complex

  p53 gene

  apoptosis

  tumor suppressor genes

  MPF

   prophase

   prometaphase

metaphase

anaphase

telophase

Meiosis

  reductional

  equational

  prophase I

   leptonema

    homology search

    rough pairing

   zygonema

   pachynema

    synapsis

   diplonema

   diakinesis (terminalization)

  metaphase I

anaphase I

  disjunction, nondisjunction

telophase I

  random segregation

  independent assortment

Meiosis II

  prophase II

  metaphase II

  anaphase II

  telophase II

Chapter 8 Mitosis and Meiosis

Spermatogenesis

Spermiogenesis

Oogenesis

First meiotic division

Second meiotic division

Nondisjunction

Sexual reproduction

    reshuffle chromosomes

    provides for crossing over and
    variation

Folded-fiber model

Concepts

Prokaryotic

Eukaryotic

Endosymbiont hypothesis

Homologous chromosomes (F8.2)

    diploid number (2n) (F8.1)

    loci, locus (F8.2)

    haploid genome (haploid number, n)

        biparental inheritance

        alleles (F8.2)

Cell cycle regulation

    tumor suppression

Crossing over

Mitosis (F8.3)

    identical daughters

    equivalent genetic information

Meiosis (F8.4)

    produces haploid gametes or spores

    reshuffles genetic combinations
    (chromosomes)

    genetic recombination

        (crossing over)

    constant amount of genetic material

    production of variation

Nondisjunction

Fertilization

    reconstitution of genetic material

Segregation

Independent assortment

Primary nondisjunction (meiosis I)

Secondary nondisjunction (meiosis II)

Gametophyte stage

Sporophyte stage

Sexual reproduction

F8.1 Diagram illustrating relationships among stages of interphase. Also illustrated are chromosomes, chromosome number and structure in an organism with a diploid chromosome number of 4 (2*n* = 4). There are two pairs of chromosomes, one large metacentric, one smaller metacentric. Individual chromosomes cannot be seen at interphase, therefore the chromosomes pictured here are hypothetical. Notice that in mitosis there is no change in chromosome number even though the DNA content doubles during the S phase. The chromosomes become doubled structures as a result of S phase.

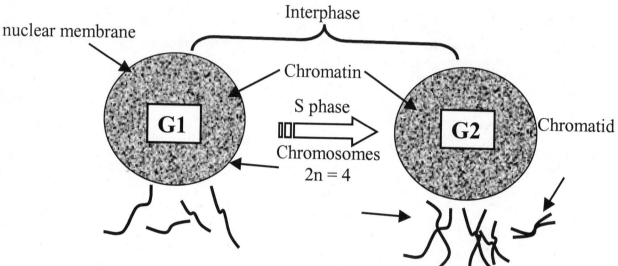

F8.2 Important nomenclature referring to chromosomes and genes in an organism where the diploid chromosome number is 4 (2*n* = 4). There are two pairs of chromosomes, one large metacentric, one small telocentric. Sister chromatids are identical to each other while homologous chromosomes are similar to each other in terms of overall size, centromere location, function, and other factors described in the text.

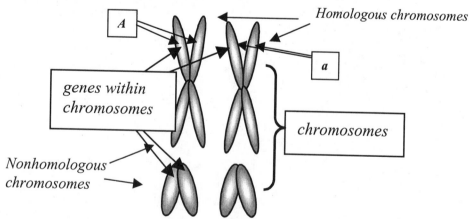

F8.3 Illustration of chromosomes of mitotic cells in an organism with a chromosome number of 4 (2*n* = 4).

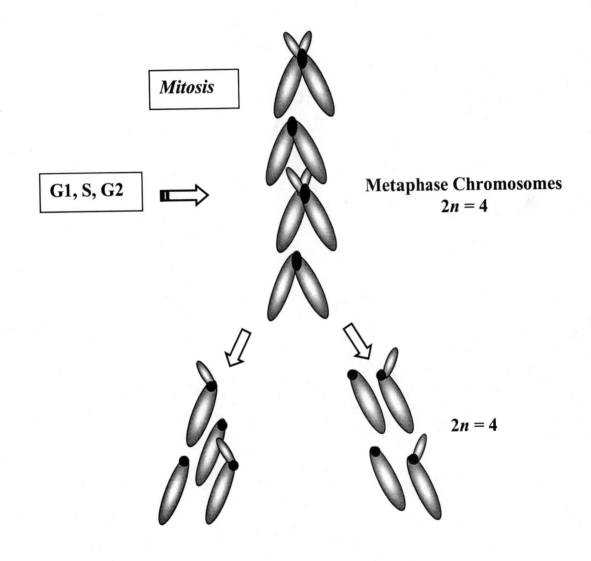

F8.4 Illustration of chromosomes of meiotic cells in an organism with a chromosome number of 4 (2n = 4).

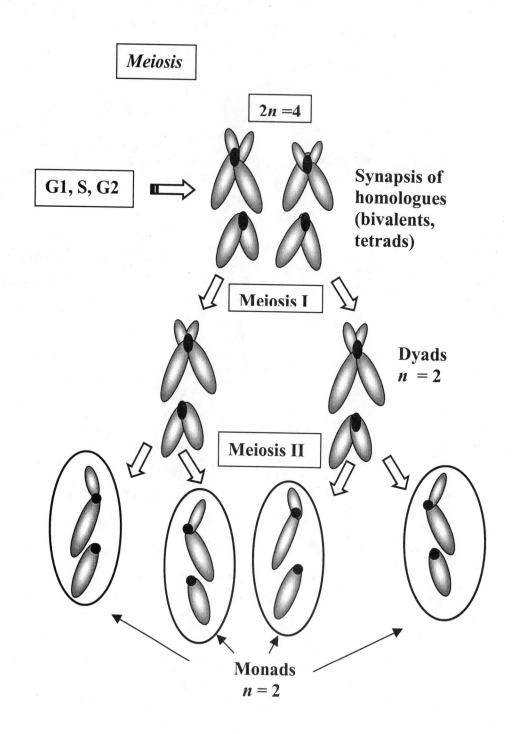

Solutions to Problems and Discussion Questions

1. (a) During interphase of the cell cycle (mitotic and meiotic), chromosomes are not condensed and are in a genetically active, spread out, form. In this condition, chromosomes are not visible as individual structures under the microscope (light or electron). See F8.1 for a sketch of what *chromatin* might look like. Chromatin contains the genetic material which is responsible for maintaining hereditary information (from one cell to daughter cells and from one generation to the next) and production of the phenotype.

(b) The *nucleolus (pl. nucleoli)* is a structure which is produced by activity of the nucleolar organizer region in eukaryotes. Composed of ribosomal RNA and protein, it is the structure for the production of ribosomes. Some nuclei have more than one *nucleolus*. Nucleoli are not present during mitosis or meiosis because in the condensed state of chromosomes, there is little or no RNA synthesis.

(c) The *ribosome* is the structure where various RNAs, enzymes, and other molecular species assemble the primary sequence of a protein. That is, amino acids are placed in order as specified by messenger RNA. Ribosomes are relatively non-specific in that virtually any ribosome can be used in the translation of any mRNA. The structure and function of the ribosome will be described in greater detail in later chapters of the K/C text.

(d) The *mitochondrion (pl. mitochondria)* is a membrane-bound structure located in the cytoplasm of eukaryotic cells. It is the site of oxidative phosphorylation and production of relatively large amounts of ATP. It is the trapping of energy in ATP which drives many important metabolic processes in living systems.

(e) The *centriole* is a cytoplasmic structure involved (through the formation of spindle fibers) in the migration of chromosomes during mitosis and meiosis.

(f) The *centromere* serves as an attachment point for sister chromatids (see F8.1) and a region where spindle fibers attach to chromosomes (kinetochore). The centromere divides during mitosis and meiosis II thus aiding in the partitioning of chromosomal material to daughter cells. Failure of centromeres or spindle fibers to function properly may result in nondisjunction.

2. One of the most important concepts to be gained from this chapter is the relationship which exists among chromosomes in a single cell. Chromosomes which are homologous share many properties including:

Overall length: look carefully at F8.1 and F8.2, to see that each cell prior to anaphase I, contains two chromosomes of approximately the same overall length.

Position of the centromere (metacentric, submetacentric, acrocentric, telocentric); again, look carefully at F8.1 and F8.2. Notice that in each if there is one metacentric chromosome, there will be another metacentric chromosome.

Banding patterns: look carefully at the K/C text and notice how similar the banding patterns are of the homologous (side-by-side) chromosomes. Look at the two homologous chromosomes of pair #1 for example. While the overall length of chromosome pairs #16 and #17 appear to be the same note the difference in banding patterns of these non-homologous chromosomes. Compare chromosome pairs #19 and #20 for a more striking example.

Chapter 8 Mitosis and Meiosis

Notice also that sister chromatids have identical banding patterns as would be expected since sister chromatids are, with the exception of mutation, identical copies of each other. We would expect that homologous chromosomes would have banding patterns which are very similar (but not identical) because homologous chromosomes are genetically similar but not genetically identical.

Type and location of genes: notice in F8.2 that a locus signifies the location of a gene along a chromosome. What that really means is that for each characteristic specified by a gene, like blood type, eye color, skin pigmentation, there are genes located along chromosomes. The *order* of such loci is identical in homologous chromosomes, but the genes themselves, while being in the same order, may not be identical. Look carefully at the inset (box) in the upper portion of F8.2 and see that there are alternative forms of genes, *A* and *a*, at the same location along the chromosome. *A* and *a* are located at the same place and specify the same *characteristic* (eye color, for example) but there are slightly different manifestations of eye color (*brown* vs. *blue* for example). Just as an individual may inherit gene *A* from the father and gene *a* from the mother, each zygote inherits one homologue of each pair from the father and one homologue of each pair from the mother.

Autoradiographic pattern; homologous chromosomes tend to replicate during the same time of S phase.

Diploidy is a term often used in conjunction with the symbol $2n$. It means that both members of a homologous pair of chromosomes are present. Refer to F8.1 in this book. Notice that during mitosis, the normal chromosome complement is $2n$ or diploid. In humans, the diploid chromosome number is 46 while in *Drosophila melanogaster* it is 8. The K/C text lists the *haploid* chromosome number for a variety of species.

Notice that in man and flies, the haploid chromosome number is one-half the diploid number. This applies to other organisms as well. However, it is very important to realize that *haploidy* specifically refers to the fact that each haploid cell contains *one chromosome of each homologous pair of chromosomes*.

Compare the nuclear contents of a spermatid and a cell at zygonema in the K/C text. Note that each spermatid contains one member of each of the original chromosome pairs (seen at zygonema). Haploidy is usually symbolized as *n*.

The change from a diploid ($2n$) to haploid (n) occurs during *reduction division* when tetrads become dyads during meiosis I. Referring to the number of human chromosomes, the primary spermatocyte ($2n = 46$) becomes two secondary spermatocytes each with $n = 23$.

3. As you examine the criteria for *homology* in question #2 above, you can see that overall length and centromere position are but two factors required for homology. Most importantly, genetic content in non-homologous chromosomes is expected to be quite different. Other factors including banding pattern and time of replication during S phase would also be expected to vary among non-homologous chromosomes.

4. Because a section of Chapter #8 deals with mitosis, it would be best to deal with this question by reading the appropriate section in the K/C text and examining corresponding figures. Understanding mitosis and all the related terms is essential for an understanding of genetics. There are several sample test questions at the end of this book which will help you determine your understanding of mitosis.

5. The first sentence tells you that $2n = 16$ and it is a question about mitosis. Since each chromosome in prophase is doubled (having gone through an S phase) and is visible at the end of prophase, there should be 32 chromatids. Because the centromeres divide and what were previously sister chromatids migrate to opposite poles during anaphase, there should be 16 chromosomes moving to each pole. If you refer to F8.1 you will see an example with $2n = 4$ and that there are four doubled chromosomes in prophase. Notice that there are eight chromatids visible at late prophase.

6. Refer to K/C figures for an explanation. Notice the different anaphase shapes of chromosomes as they are moving to the poles. Your understanding of these structures will be determined by several of the sample test questions at the end of this book.

7. Because of a cell wall around the plasma membrane in plants, a cell plate, which was laid down during anaphase, becomes the middle lamella where primary and secondary layers of the cell wall are deposited.

8. Carefully read the section on mitosis and cell division in the text. Refer to F8.1 for information pertaining to the interphase. Refer to K/C figures for a diagram of mitosis. Notice that, in contrast to meiosis, there is no pairing of homologous chromosomes in mitosis and the chromosome number does not change. Regulatory checkpoints include those in G1/S, G2/M, and mitosis. Each allows checks on intracellular and environmental conditions.

Autoradiography is a technique which can be used to determine that there is DNA synthesis during the interphase. If DNA precursors are available early in interphase, they can be incorporated into replicating DNA during the S phase. Autoradiographs made after the S phase will show radioactive material in the DNA. Complexed cdk and cyclin proteins provide the mechanism for passage from one stage of the cell cycle to the next.

9. Not necessarily. If crossing over occurred in meiosis I, then the chromatids in the secondary oocyte are not identical. Once they separate during meiosis II, unlike chromatids reside in the ootid and the second polar body.

10. Compared with mitosis which maintains a chromosomal constancy, meiosis provides for a reduction in chromosome number, and an opportunity for exchange of genetic material between homologous chromosomes. In mitosis there is no change in chromosome number or kind in the two daughter cells whereas in meiosis numerous potentially different haploid (*n*) cells are produced. During oogenesis, only one of the four meiotic products is functional; however, four of the four meiotic products of spermatogenesis are potentially functional.

11. (a) *Synapsis* is the point-by-point pairing of homologous chromosomes during prophase of meiosis I.

(b) *Bivalents* are those structures formed by the synapsis of homologous chromosomes. In other words, there are two chromosomes (and four chromatids) which make up a bivalent. If an organism has a diploid chromosome number of 46, then there will be 23 bivalents in meiosis I.

(c) *Chiasmata* is the plural form of chiasma and refers to the structure, when viewed microscopically, of crossed chromatids. Notice in K/C figures in the text, the exchange of chromatid pieces in diplonema and diakinesis.

(d) *Crossing over* is the exchange of genetic material between chromatids. Also called recombination, it is a method of providing genetic variation through the breaking and rejoining of chromatids.

(e) *Chromomeres* are bands of chromatin which look different from neighboring patches along the length of a chromosome.

(f) Examine F8.1 in this book. Notice that *sister chromatids* are "post-S phase" structures of replicated chromosomes. Sister chromatids are genetically identical (except where mutations have occurred) and are originally attached to the same centromere. Identify sister chromatids in the figures in the text. Note that sister chromatids separate from each other during anaphase of mitosis and anaphase II of meiosis.

(g) Tetrads are synapsed homologous chromosomes thereby composed of four chromatids. There are as many tetrads as the haploid chromosome number.

(h) Actually, each tetrad is made of two dyads which separate from each other during anaphase I of meiosis. Note that dyads are composed of two chromatids joined by a centromere.

(i) At anaphase II of meiosis, the centromeres divide and sister chromatids go to opposite poles.

(j) The *synaptonemal complex* is a nuclear structure which is often found associated with synapsed meiotic chromosomes. It looks and acts like a zipper.

12. Sister chromatids are genetically identical, except where mutations may have occurred during DNA replication. Nonsister chromatids are genetically similar if on homologous chromosomes or genetically dissimilar if on nonhomologous chromosomes. If crossing over occurs, then chromatids attached to the same centromere will no longer be identical.

13. During meiosis I chromosome number is reduced to haploid complements. This is achieved by synapsis of homologous chromosomes and their subsequent separation. It would seem to be more mechanically difficult for genetically identical daughters to form from mitosis if homologous chromosomes paired.

14. Look carefully at F8.4 in this book and notice that for a cell with 4 chromosomes, there are two tetrads each comprised of a homologous pair of chromosomes.

(a) If there are 16 chromosomes there should be 8 tetrads.
(b) Also note that, after meiosis I and in the second meiotic prophase, there are as many dyads as there are pairs of chromosomes. There will be 8 dyads.
(c) Because the monads migrate to opposite poles during meiosis II (from the separation of dyads) there should be 8 monads migrating to *each* pole.
(d) $(1/2)^8$

15. Examine appropriate figures in the K/C text. Notice that major differences include the sex in which each occurs, and that the distribution of cytoplasm is unequal in oogenesis but considered to be equal in the products of spermatogenesis. Chromosomal behavior is the same in spermatogenesis and oogenesis except that the nuclear activity in oogenesis is "off-center" thereby producing first and second polar bodies by unequal cytoplasmic division. Each spermatogonium and primary spermatocyte produces four spermatids whereas each oogonium and primary oocyte produces one ootid. Because early development occurs in the absence of outside nutrients, it is likely that the unequal distribution of cytoplasm in oogenesis evolved to provide sufficient information and nutrients to support development until the transcriptional activities of the zygotic nucleus begin to provide products. Polar bodies probably represent non-functional by-products of such evolution.

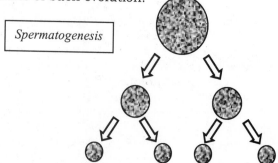

Spermatogenesis

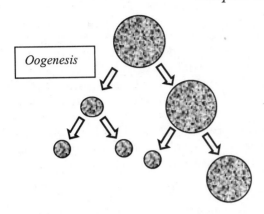

Oogenesis

16. This answer contains several parts. First, through independent assortment of chromosomes at anaphase I of meiosis, daughter cells (secondary spermatocytes and secondary oocytes) may contain different sets of maternally- and paternally-derived chromosomes. Examine the diagram below. Notice that there are several ways in which the maternally- and paternally-derived chromosomes may align. Can you calculate the probability of all the maternally-derived chromosomes going to the "right-hand" pole? Second, crossing over, which happens at a much higher frequency in meiotic cells as compared to mitotic cells, allows maternally- and paternally-derived chromosomes to exchange segments thereby increasing the likelihood that daughter cells (that is, secondary spermatocytes and secondary oocytes) are genetically unique.

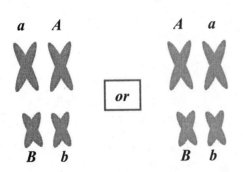

Notice that there are two different orientations of tetrads in meiosis. Independent assortment of nonhomologous chromosomes adds to genetic variability. Daughter cells resulting from the process of mitosis are usually genetically identical.

17. As you first read this question, think about an animal with $n = 6$, therefore there will be 6 tetrads. The question concerns one of these tetrads as it passes normally through meiosis I, but one dyad undergoes secondary nondisjunction. Secondary nondisjunction occurs during the second meiotic division.

(a) The mature ovum should contain $n+1$ chromosomes: the five chromosomes from normal disjunction and two (from one dyad) from the nondisjunctional chromosome.

(b) The second polar body did not receive one of the six monads it would normally receive, so it should have five monads (which are chromosomes).

(c) When the normal sperm with its n chromosome number combines with an $n+1$ ovum, it will produce a zygote with $2n+1$, or thirteen chromosomes. This condition is termed, *trisomy*.

18. One half of each tetrad will have a maternal homolog: $(1/2)^{10}$.

19. (a) While it is likely that the molecular processes involved in crossing over occur earlier, crossing over is known to have occurred by pachynema.

(b) *Synapsis* begins at zygonema with continuation of the homology search and when homologous chromosomes align in a point-by-point fashion to form bivalents or tetrads. More intimate pairing (synapsis) is completed during pachynema.

(c) Chromosomes begin to condense at the earliest stage of prophase I, leptonema.

(d) *Chiasmata* are clearly visible at diplonema.

20. In angiosperms, meiosis results in the formation of microspores (male) and megaspores (female) which give rise to the haploid male and female gametophyte stage. Micro- and megagametophytes produce the pollen and the ovules respectively. Following fertilization, the sporophyte is formed.

21. The transition from chromatin to individual chromosomes occurs at the beginning of mitosis (or meiosis). During this time, chromatin fibers fold up and condense into the typical mitotic chromosome. The *folded-fiber model* depicts this transition.

22. The transition is at the end of interphase and the beginning of mitosis (prophase) when the chromosomes are in the condensation process. This eventually leads to the typically shortened and "fattened" metaphase chromosome.

23. There are three primary checkpoints in the cell cycle. The G1/S checkpoint monitors cell size and DNA damage. The G2/M checkpoint monitors physiological conditions in the cell and determines the extent of DNA replication and repair. The M checkpoint monitors chromosome attachment and positioning. Failures at any one of these checkpoints can lead to uncontrolled cell proliferation and cancer.

24. Mutations symbolized as *cdc* are those involved in regulating the cell division cycle. Through the use of such mutations, kinases and their involvement with cyclins were also discovered. In addition, *cdc* mutations helped in the discovery of cell cycle checkpoints.

25. *p53* plays a role in regulating the G1 to S transition. It is a tumor suppressor. Mutations cause a lack of control over the cell cycle and cancer often follows.

26. They would probably be homologous chromosomes and contain similar (but not identical) genetic information. Their centromeres would most likely be in the same position relative to chromosome arm lengths and any physical characteristics such as secondary constrictions or bands would be similar. They would have a similar sequence of nitrogenous bases. They would most likely replicate synchronously during the S phase of the cell cycle.

27. (a) Duplicated chromosomes A^m, A^p, B^m, B^p, C^m, and C^p will align at metaphase, with the centromeres dividing and sister chromatids going to opposite poles at anaphase.

(b) Side-by-side alignment of A^m, A^p, B^m B^p, C^m, C^p will occur in various arrangements.

(c,d) Eight possible combinations of meiotic II products will occur: A^m, B^p, C^m, for example (each with sister chromatids).

(e) After meiosis I, the two product cells would be as follows: A^m or A^p, B^m or B^p, C^m and C^p; A^m or A^p, B^m or B^p, no C^m or C^p.

(f) Two products will be trisomic for chromosome C and disomic for chromosomes A and B. Two products will be monosomic for chromosome C and disomic for chromosomes A and B.

Chapter 9: Mendelian Genetics

Concept Areas	Corresponding Problems
Mendel's Model	3, 5, 11, 12, 13, 14, 16
Monohybrid Crosses	1, 2, 4, 6, 18, 19, 20, 21, 22, 26, 27, 34, 38
Dihybrid Crosses	7, 8, 9, 10, 15, 23, 32, 39
Trihybrid Crosses	17, 28, 29, 30, 37
Independent Assortment	16, 28
Probability	31, 33, 34, 35, 36, 38, 40
Chi-Square Analysis	24, 25, 41
Pedigree Analysis	26, 27, 33

Vocabulary: Organization and Listing of Terms and Concepts

Historical

Mendelian Genetics (Gregor Mendel)

  Pisum sativum (1865)

    units of heredity (particulate)

  Rebirth (1900)

    Hugo DeVries

    Karl Correns

    Erich Tschermak

Punnett squares

Pascal's triangle

Structures and Substances

Unit factors, Genes

Alleles

Processes/Methods

Transmission genetics

  true-breeding, "bred true"

monohybrid cross

selfing, self-fertilizing

reciprocal cross

test cross

parental generation (P_1)

first filial generation (F_1)

second filial generation (F_2)

ratios

  3:1, 1:1

  1:2:1

  9:3:3:1, 1:1:1:1

  27:9:9:9:3:3:3:1

product law

  2^n (n = haploid chromosome number)

sum law

  conditional probability

Chapter 9 Mendelian Genetics

binomial theorem

$(a + b)^n$

dihybrid cross (two-factor cross)

trihybrid cross (three-factor cross)

forked-line (branch diagram) method

Statistical testing (analysis)

predicted occurrences

proportions

sample size

chance deviation

random fluctuations

null hypothesis

measured (observed) values

predicted values

goodness of fit

chi-square analysis (χ^2)

degrees of freedom

probability value (p)

reject the null hypothesis

fail to reject the null hypothesis

0.05 probability value

Pedigree

sibs, sibship line

monozygotic (identical) twins

dizygotic (fraternal) twins

propositus, propositi

Concepts

Unit factors in pairs (F9.1)

Dominance/recessiveness (F9.1)

Symbolism (F9.2, F9.3)

Segregation

Phenotype

Genotype

Homozygous (homozygote)

Heterozygous (heterozygote)

Independent assortment

Genotypic ratio

Variation of a continuous nature

Variation of a discontinuous nature

Chromosome theory of heredity

Diploid number

Probability

Statistical testing

F9.1 Illustration of the union of maternal and paternal genes (*A* and *a*) to give two genes in the zygote. Mendelian "unit factors" occur in pairs in diploid organisms. Dominant genes are often given the upper case letter as the symbol while the lower case letter is often used to symbolize the recessive gene.

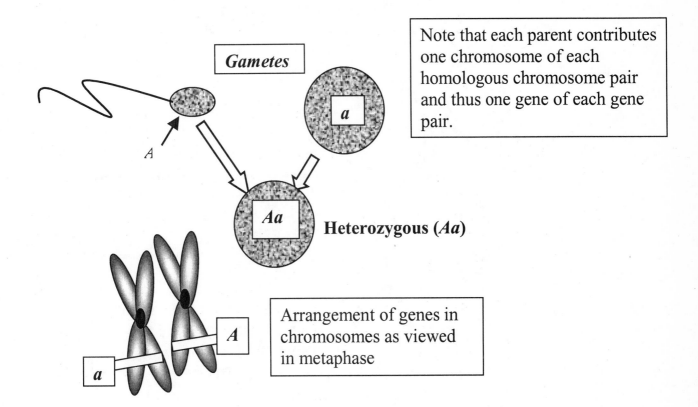

Gametes

Note that each parent contributes one chromosome of each homologous chromosome pair and thus one gene of each gene pair.

A

a

Aa **Heterozygous (*Aa*)**

Arrangement of genes in chromosomes as viewed in metaphase

A

a

F9.2 Critical symbolism associated with genes and chromosomes. Below are positioned two different gene pairs (*Aa* and *Bb*) on non-homologous chromosomes. Note, with two different gene pairs two different characteristics may be involved such as seed shape (*A* and *a*) and seed color (*B* and *b*).

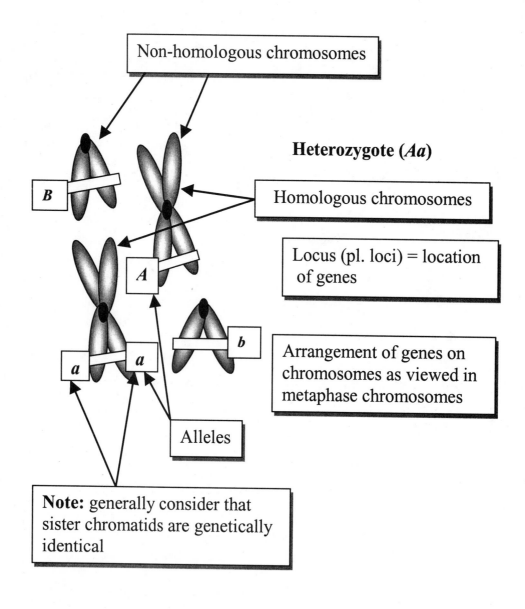

Non-homologous chromosomes

Heterozygote (*Aa*)

B

Homologous chromosomes

A

Locus (pl. loci) = location of genes

b

Arrangement of genes on chromosomes as viewed in metaphase chromosomes

a   *a*

Alleles

Note: generally consider that sister chromatids are genetically identical

F9.3 One of the most important concepts is illustrated in the figure below. Two gene pairs (*W* and *B*) are presented, each representing a different characteristic, seed shape (*W* or *w*) and seed color (*B* or *b*). Different gene pairs may influence completely different characteristics (as indicated here) or the same characteristics (described in Chapter 4).

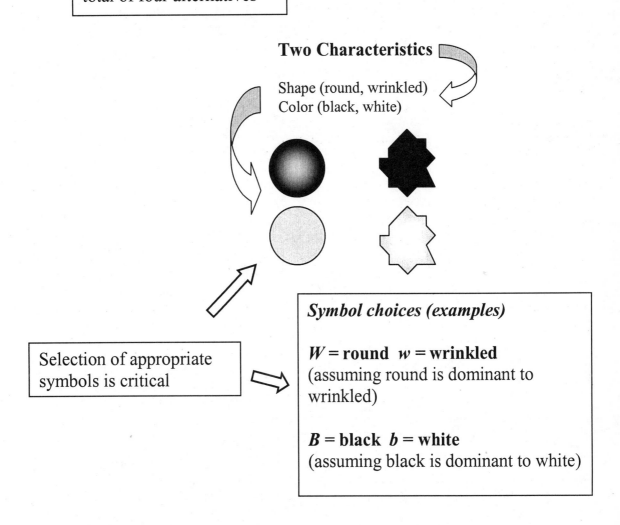

Two characteristics but a total of four alternatives

Two Characteristics

Shape (round, wrinkled)
Color (black, white)

Selection of appropriate symbols is critical

Symbol choices (examples)

W = **round** *w* = **wrinkled**
(assuming round is dominant to wrinkled)

B = **black** *b* = **white**
(assuming black is dominant to white)

Chapter 9 Mendelian Genetics

Solutions to Problems and Discussion Questions

1. Several points surface in the first sentence of this question. First, two alternatives (black and white) of one characteristic (coat color) are being described, therefore a monohybrid condition exists.

Second, are the guinea pigs in the parental generation (P_1) homozygous or heterozygous? Notice in the introductory sentence, just after PROBLEMS AND DISCUSSION QUESTIONS, there is the statement "members of the P_1 generation are homozygous. . ."

Third, which is dominant, *black* or *white* ? Note that all the offspring are black, therefore black can be considered dominant. The second sentence of the problem verifies that a monohybrid cross is involved because of the 3/4 black and 1/4 white distribution in the offspring. Referring to appropriate K/C figures and knowing that genes occur in pairs in diploid organisms, one can write the genotypes and the phenotypes requested in part (a) as follows:

(a)

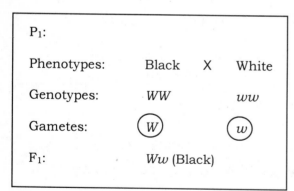

F_1 X F_1:

| Phenotypes: | Black | X | Black |
| Genotypes: | *Ww* | | *Ww* |

Gametes: (W) (w) (W) (w)

(combine as in the K/C text)

F_2:

Phenotypes: Black Black Black White

Genotypes: *WW* *Ww* *Ww* *ww*

(b) Since *white* is a recessive gene (to *black*) each white guinea pig must be homozygous and a cross between two white guinea pigs must produce all white offspring.

| white | X | white |
| *ww* | | *ww* |

(c) Recall the various possibilities of the genotypes capable of producing the black phenotype in the F_2 generation in part (a) above: *WW* and *Ww*. In Cross 1 in the problem, all black offspring are observed and the most likely parental genotypes would be as follows:

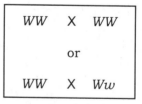

There is the possibility that black guinea pigs of the *Ww* genotype could produce all black offspring if the sample size was such that *ww* offspring were not produced. In Cross 2, a typical 3:1 Mendelian ratio is observed which indicates that two heterozygotes were crossed:

| *Ww* | X | *Ww* |

2. Start out with the following gene symbols;

    A   = normal (not albino),

    a   = albino.

Since albinism is inherited as a recessive trait, genotypes *AA* and *Aa* should produce the normal phenotype, while *aa* will give albinism.

(a) The parents are both normal, therefore they could be either *AA* or *Aa*. The fact that they produce an albino child requires that each parent provides an *a* gene to the albino child; thus the parents must both be heterozygous (*Aa*).

(b) To start out, the normal male could have either the *AA* or *Aa* genotype. The female must be *aa*. Since all the children are normal one would consider the male to be *AA* instead of *Aa* . However, the male could be *Aa*. Under that circumstance, the likelihood of having six children, all normal is 1/64.

(c) To start out, the normal male could have either the *AA* or *Aa* genotype. The female must be *aa*. The fact that half of the children are normal and half are albino indicates a typical "test cross" in which the *Aa* male is mated to the aa female.

(d)

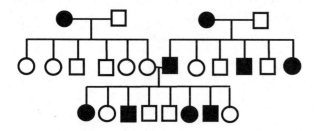

The 1:1 ratio of albino to normal in the last generation theoretically results because the mother is *Aa* and the father is *aa*.

3. *Unit Factors in Pairs*: It is important to see that each time a phenotype (normal or abnormal) is being stated, genotypes are symbolized as pairs of genes; *AA*, *Aa* or *aa*. Review F9.2 to understand the need to assign appropriate symbols to genes.

Dominance and Recessiveness: Because the gene for normal pigmentation is completely dominant over the gene for albinism (*a* is fully recessive), it was necessary to consider, at first, whether normally pigmented individuals in the problem were homozygous normal (*AA*) or heterozygous (*Aa*). By looking at the frequency of expression of the recessive gene in the offspring (in *aa* individuals) one can often determine an *Aa* type from an *AA* type.

Segregation: During gamete formation when homologous chromosomes move to opposite poles, paired elements (genes) separate from each other.

4. While it is very difficult, if not impossible, to know exactly how Mendel made the step from his "monohybrid results" to his postulates, he was able to develop a model with several important components or postulates.

First, organisms contained **unit factors** for various traits.

Second, if these **factors occurred in pairs**, there existed the possibility that some organisms would "breed true" if homozygous, while others would not (heterozygotes). If one "factor" of a pair had a **dominant influence** over the other, then he could explain how two organisms, looking the same, could be genetically different (homozygous or heterozygous).

Third, if the **paired elements separate (segregate)** from each other during gamete formation and if gametes combine at random, he could account for the 3:1 ratios in the monohybrid crosses.

Three excellent books give insight into Mendel's life and the context of his discoveries: Carlson, E. A. 1966. *The Gene: A Critical History*. Philadelphia: W. B. Saunders.; Sturtevant, A. H. 1965. A History of Genetics. New York: Harper and Row.; Voeller, B. R. 1968. *The Chromosome Theory of Inheritance*. New York: Appleton-Century-Crofts.

5. *Pisum sativum* is easy to cultivate. It is naturally self-fertilizing, but it can be crossbred. It has several visible features (*e.g.*, tall or short, red flowers or white flowers) which are consistent under a variety of environmental conditions yet contrast due to genetic circumstances. Seeds could be obtained from local merchants.

6. With any "long" and involved problem, students often have trouble getting started in the right direction, and seeing the problem through to the necessary conclusions. First, read the entire question and see that you are to determine (1) the pattern of inheritance for "checkered and plain," and (2) the gene symbols and genotypes of all the parents and offspring. Notice that there is reference to one characteristic, *pattern*, with two alternatives, checkered vs. plain.

We should consider this to be a monohybrid condition unless complications arise. We are to use the respective offspring from the P_1 cross (F_1 progeny a, b, and c) in a series of F_1 crosses, d through g. Approach the problem by first assigning *probable* genotypes to the P_1 crosses then, where there is ambiguity [such as cross (a)], use the F_1 crosses for clarification.

Assignment of symbols:

P = checkered; p = plain. Checkered is tentatively assigned the dominant function because in a casual examination of the data, especially cross (b), we see that checkered types are more likely to be produced than plain types.

Cross (a):

PP X PP or PP X Pp

Notice in cross (d) that the checkered offspring, when crossed to plain, produce only checkered F_2 progeny and in cross (g) when crossed to checkered still produce only checkered progeny. From this additional information, one can conclude that in the progeny of cross (a) there are no heterozygotes and the original cross must have been PP X PP.

Cross (b):

PP X pp

This assignment seems reasonable because among 38 offspring, no plain types are produced. In addition, we would expect all the F_1 progeny to be heterozygous and if crossed to plain, as in cross (e), to produce approximately half checkered and half plain offspring. In cross (f) we would expect such heterozygotes to produce a 3:1 ratio, which is observed.

Cross (c):

Because all the offspring from this cross are plain, there is no doubt that the genotype of both parents is *pp*.

Genotypes of all individuals:

P₁ Cross	F₁ Progeny Checkered	Plain
(a) *PP* X *PP*	*PP*	
(b) *PP* X *pp*	*Pp*	
(c) *pp* X *pp*		*pp*
(d) *PP* X *pp*	*Pp*	
(e) *Pp* X *pp*	*Pp*	*pp*
(f) *Pp* X *Pp*	*PP, Pp*	*pp*
(g) *PP* X *Pp*	*PP, Pp*	

7. In the first sentence you are told that there are two *characteristics* which are being studied; seed shape and cotyledon color. Expect, therefore, this to be a dihybrid situation with *two gene pairs* involved. One also sees the possible alternatives of these two characteristics: *seed shape* ; wrinkled vs. round; *cotyledon color* ; green vs. yellow. After reading the second sentence you can predict that the gene for round seeds is dominant to that for wrinkled seeds and the gene for yellow cotyledons is dominant to the gene for green cotyledons.

Symbolism:

w = wrinkled seeds   *g* = green cotyledons

W = round seeds   *G* = yellow cotyledons

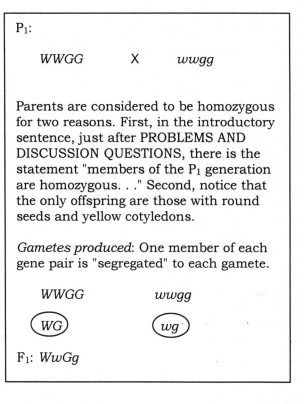

P₁:

      WWGG   X   *wwgg*

Parents are considered to be homozygous for two reasons. First, in the introductory sentence, just after PROBLEMS AND DISCUSSION QUESTIONS, there is the statement "members of the P₁ generation are homozygous. . ." Second, notice that the only offspring are those with round seeds and yellow cotyledons.

Gametes produced: One member of each gene pair is "segregated" to each gamete.

     WWGG   *wwgg*

     (*WG*)   (*wg*)

F₁: *WwGg*

F₁ X F₁:

     WwGg X *WwGg*

Gametes produced: Under conditions of independent assortment, there will be four (2^n, where n = number of heterozygous gene pairs) different types of gametes produced by each parent.

Punnett Square:

	WG	*Wg*	*wG*	*wg*
WG	*WWGG*	*WWGg*	*WwGG*	*WwGg*
Wg	*WWGg*	*WWgg*	*WwGg*	*Wwgg*
wG	*WwGG*	*WwGg*	*wwGG*	*wwGg*
wg	*WwGg*	*Wwgg*	*wwGg*	*wwgg*

Collecting the phenotypes according the dominance scheme presented above, gives the following:

9/16 *W_G_* round seeds, yellow cotyledons

3/16 *W_gg* round seeds, green cotyledons

3/16 *wwG_* wrinkled seeds, yellow cotyledons

1/16 *wwgg* wrinkled seeds, green cotyledons

Notice that a dash (_) is used where, because of dominance, it makes no difference as to the dominant/recessive status of the allele.

Forked, or branch diagram:

Seed shape	*Cotyledon color*	*Phenotypes*

```
            3/4 yellow ⟹ 9/16 round, yellow
           /
  3/4 round
           \
            1/4 green  ⟹ 3/16 round, green

            3/4 yellow ⟹ 3/16 wrinkled, yellow
           /
  1/4 wrinkled
           \
            1/4 green  ⟹ 1/16 wrinkled, green
```

8. *WWgg* = 1/16

9. Symbolism as before:

w = wrinkled seeds	*g* = green cotyledons
W = round seeds	*G* = yellow cotyledons

Examine each characteristic (seed shape vs. cotyledon color) separately.

(a) Notice a 3:1 ratio for seed shape, therefore *Ww* X *Ww*; and no green cotyledons, therefore *GG* X *GG* or *GG* X *Gg*. Putting the two characteristics together gives

$$WwGG \quad X \quad WwGG$$

or

$$WwGG \quad X \quad WwGg.$$

(b) Notice a 1:1 ratio for seed shape (8/16 wrinkled and 8/16 round) and a 3:1 ratio for cotyledon color (12/16 yellow and 4/16 green). Therefore the answer is

$$wwGg \quad X \quad WwGg.$$

(c) The offspring occur in a typical 9:3:3:1 ratio, therefore the F_2 plants have the doubly heterozygous genotypes of

$$WwGg \quad X \quad WwGg.$$

(d) This is a typical 1:1:1:1 test cross (or backcross) ratio and will signify that one parent is doubly heterozygous while the other is fully homozygous recessive. The answer is

$$WwGg \quad X \quad wwgg.$$

10. A test cross involves a cross of an organism with an unknown genotype to a fully homozygous recessive organism. In Problem #9, (d) fits this description.

11. Because independent assortment may be defined as one gene pair segregating independently of another gene pair, one would need at least two gene pairs in order to demonstrate independent assortment.

12. Mendel's four postulates are related to the diagram below.

1. Factors occur in pairs. Notice *A* and *a*.
2. Some genes have dominant and recessive alleles. Notice *A* and *a*.
3. Alleles segregate from each other during gamete formation. When homologous chromosomes separate from each other at anaphase I, alleles will go to opposite poles of the meiotic apparatus.
4. One gene pair separates independently from other gene pairs. Different gene pairs on the same homologous pair of chromosomes (if far apart) or on non-homologous chromosomes will separate independently from each other during meiosis.

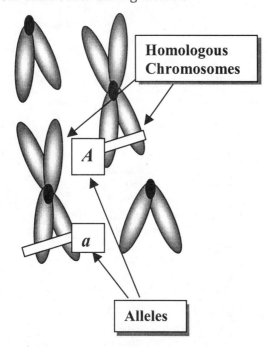

Homologous Chromosomes

A

a

Alleles

13. Carefully re-read the answer to Question #2 in Chapter #2. Briefly, the factors which specify chromosomal homology are the following:

> *overall length,*
> *position of the centromere,*
> *banding patterns,*
> *type and location of genes*,*
> *autoradiographic pattern*

*functional basis for homologous chromosome designation

14. Homozygosity refers to a condition where both genes of a pair are the same (*i.e. AA* or *GG* or *hh*) whereas heterozygosity refers to the condition where members of a gene pair are different (*i.e. Aa* or *Gg or Bb*). Homozygotes produce only one type of gamete whereas heterozygotes will produce 2^n types of gametes where n = number of heterozygous gene pairs (assuming independent assortment).

15. There are two characteristics presented here, body color and wing length. First, assign meaningful gene symbols.

Body color	Wing length
E = gray body color	*V* = long wings
e = ebony body color	*v* = vestigial wings

(a)

P$_1$:

EEVV X *eevv*

F$_1$: *EeVv* (gray, long)

F$_2$: This will be the result of a Punnett Square with 16 boxes as in the K/C text.

Phenotypes	Ratio	Genotypes	Ratio
gray, long	9/16	*EEVV*	1/16
		EEVv	2/16
		EeVV	2/16
		EeVv	4/16
gray, vestigial	3/16	*EEvv*	1/16
		Eevv	2/16
ebony, long	3/16	*eeVV*	1/16
		eeVv	2/16
ebony, vestigial	1/16	*eevv*	1/16

(b)

P$_1$:

    EEvv X *eeVV*

F$_1$: It is important to see that the results from this cross will be exactly the same as those in part (a) above. The only difference is that the recessive genes are coming from both parents, rather than from one parent only as in (a). The F$_2$ ratio will be the same as (a) also. When you have genes on the autosomes (not X-linked), independent assortment, complete dominance, and no gene interaction (see later) in a cross involving double heterozygotes, the offspring ratio will be 9:3:3:1.

(c)

P$_1$:

    EEVV X *EEvv*

F$_1$: *EEVv* (gray, long)

F$_2$: Notice that all the offspring will have grey bodies and you will get a 3:1 ratio of long to vestigial wings. You should see this before you even begin working through the problem. Even though this cross involves two gene pairs it will give a "monohybrid" type of ratio because one of the gene pairs is homozygous (body color) and **one** gene pair is heterozygous (wing length).

Phenotypes	Ratio	Genotypes	Ratio
gray, long	3/4	*EEVV*	1/4
		EEVv	2/4
gray, vestigial	1/4	*EEvv*	1/4

NOTE: After working through this problem, it is important that you try to work similar problems without constructing the time-consuming Punnett squares especially if each problem asks for phenotypic rather than genotypic ratios.

16. The general formula for determining the number of kinds of gametes produced by an organism is 2^n where n = number of *heterozygous* gene pairs.

    (a) 4: *AB, Ab, aB, ab*

    (b) 2: *AB, aB*

    (c) 8: *ABC, ABc, AbC, Abc, aBC, aBc, abC, abc*

    (d) 2: *ABc, aBc*

    (e) 4: *ABc, Abc, aBc, abc*

    (f) $2^5 = 32$

ABCDE	*aBCDE*
ABCDe	*aBCDe*
ABCdE	*aBCdE*
ABCde	*aBCde*
ABcDE	*aBcDE*
ABcDe	*aBcDe*
ABcdE	*aBcdE*
ABcde	*aBcde*
AbCDE	*abCDE*
AbCDe	*abCDe*
AbCdE	*abCdE*
AbCde	*abCde*
AbcDE	*abcDE*
AbcDe	*abcDe*
AbcdE	*abcdE*
Abcde	*abcde*

Notice that there is a pattern that can be used to write these gametes so that fewer errors will occur.

17. (a) When examining this cross

AaBbCc X AaBBCC

expect there to be eight different kinds of gametes from one parent (*AaBbCc*), and two different kinds from the other (*AaBBCC*). Therefore there should be sixteen kinds (genotypes) of offpsring (8 X 2).

Gametes: Gametes:

ABC
ABc
AbC
Abc
aBC
aBc
abC
abc

ABC
aBC

Offspring:

Genotypes Ratio   Phenotypes

AABBCC (1/16) ⎫
AABBCc (1/16) ⎪
AABbCC (1/16) ⎪
AABbCc (1/16) ⎪
AaBBCC (2/16) ⎬ A_B_C_ = 12/16
AaBBCc (2/16) ⎪
AaBbCC (2/16) ⎪
AaBbCc (2/16) ⎭

aaBBCC (1/16) ⎫
aaBBCc (1/16) ⎬ aaB_C_ = 4/16
aaBbCC (1/16) ⎪
aaBbCc (1/16) ⎭

(b) There will be four kinds of gametes for the first parent (*AaBBCc*) and two kinds of gametes for the second parent.

Gametes: Gametes:

ABC
ABc
aBC
aBc

aBC
aBc

Offspring:

Genotypes	Ratio	Phenotypes	
AaBBCC	1/8	A_BBC_	= 3/8
AaBBCc	2/8		
AaBBcc	1/8	A_BBcc	= 1/8
aaBBCC	1/8	aaBBC_	= 3/8
aaBBCc	2/8		
aaBBcc	1/8	aaBBcc	= 1/8

(c) There will be eight (2^n) different kinds of gametes from each of the parents, therefore a 64-box Punnett square. Doing this problem by the forked-line method helps considerably.

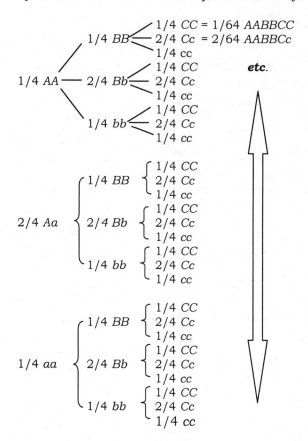

Simply multiply through each component to arrive at the final genotypic frequencies.

99

For the phenotypic frequencies, set up the problem in the following manner

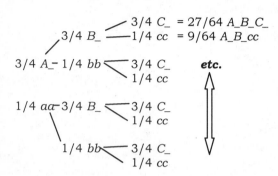

18. In the first reading of this question one should consider that there is one characteristic involved, seed color. Given that information and the F₂ progeny of 6022 yellow and 2001 green, a monohybrid condition is suspected. In addition, of the 519 self-fertilized, yellow seeded plants, 166 bred true while the others produced a 3:1 ratio of yellow to green. This is what would be expected because approximately 1/3 of the yellow F₂ plants should be homozygous, while 2/3 of the yellow F₂ plants should be heterozygous.

Set up the symbols as follows: *G* = yellow seeds, *g* = green seeds. We know that the gene for yellow seeds is dominant to that for green seeds because the F₁ is all yellow, not green.

Phenotypes	Genotypes
P₁: Yellow X green	*GG* X *gg*
F₁: all yellow	*Gg*
F₂: 6022 yellow 2001 green	1/4 *GG*; 2/4 *Gg* 1/4 *gg*

Of the yellow F₂ *offspring*, notice that 1/3 of them are *GG* and 2/3 are *Gg*. If you selfed the 1/3 *GG* types then all the offspring (the 166) would breed true whereas the others (353 which are *Gg*) should produce offspring in a 3:1 ratio when selfed.

GG   X   *GG*
= all *GG*
Gg   X   *Gg*
= 1/4 *GG*; 2/4 *Gg*; 1/4 *gg*

19. Because there is only one characteristic being dealt with in this problem (coat color) and a 3:1 ratio is mentioned in the second sentence, one can initially consider this to be a monohybrid condition. Set the gene symbols as *W* = black, *w* = white.

One hundred black animals could be all *WW*, all *Ww*, or a mixture (*WW, Ww*). When the *WW* individuals are crossed to *ww*, all the offspring will be black, which is what occurred in 94 of the cases. All these black offspring would be heterozygous and be expected to produce a 3:1 ratio when intercrossed. In those cases where a 1:1 ratio resulted, the parents must have been *Ww*. The cross would be as follows:

P₁:
Ww   X   *ww*
F₁:
1/2 *Ww*, 1/2 *ww*

If one were to cross the black and white guinea pigs from the above cross, again the offspring would produce 1/2 black and 1/2 white as above.

20. In reading this question, notice that there are two characteristics being considered; seed color (yellow, green) and seed shape (round, wrinkled). At this point you should be able to do this problem without writing down each of the steps. The F_1 can be considered to be a double heterozygote (with round and yellow being dominant). See the cross this way:

Symbols:

Seed shape	Seed color
W = round	G = yellow
w = wrinkled	g = green

P_1:   *WWgg*   X   *wwGG*

F_1:   *WwGg* cross to *wwgg*

(which is a typical testcross)

The offspring will occur in a typical 1:1:1:1 as

1/4 *WwGg* (round, yellow)

1/4 *Wwgg* (round, green)

1/4 *wwGg* (wrinkled, yellow)

1/4 *wwgg* (wrinkled, green)

Again, at this point it would be very helpful if you could do such simple problems by inspection.

21. This question deals with the definition of dominance/recessiveness. Notice that there are only two alleles (one gene pair) and three phenotypes associated with the problem; normal, "minor" anemia and "major" anemia. Under a monohybrid model, the heterozygote is distinguishable from either homozygote which does not fit the definition of dominance. One would conclude that no dominance is involved and incomplete dominance exists (see Chapter 10).

22. Since these are F_2 results from monohybrid crosses a 3:1 ratio is expected for each. Referring to the K/C text one can set up the analysis easily.

(a)

Expected ratio	Observed (o)	Expected (e)
3/4	882	885.75
1/4	299	295.25

Expected values are derived by multiplying the expected ratio by the total number of organisms.

$$\chi^2 = \Sigma \ (o - e)^2/e = .064$$

In looking in the χ^2 table with 1 degree of freedom (because there were two classes, therefore n-1 or 1 degree of freedom), we find a probability (p) value between 0.9 and 0.5.

We would therefore say that there is a "good fit" between the observed and expected values. Notice that as the deviations between the observed and expected values increase the value of χ^2 increases. So the higher the χ^2 value the more likely the null hypothesis will be rejected.

(b)

Expected ratio	Observed (o)	Expected (e)
3/4	705	696.75
1/4	224	232.25

$$\chi^2 = 0.39$$

The p value in the table for 1 degree of freedom is still between 0.9 and 0.5, however because the χ^2 value is larger in (b) we should say that the deviations from expectation are greater. The deviation in each case can be attributed to chance.

Chapter 9 Mendelian Genetics

23. One must think of this problem as a dihybrid F_2 situation with the following expectations:

Expected ratio	Observed (o)	Expected (e)
9/16	315	312.75
3/16	108	104.25
3/16	101	104.25
1/16	32	34.75

$$\chi^2 = 0.47$$

Looking at the table in the K/C text one can see that this χ^2 value is associated with a probability greater than 0.90 for 3 degrees of freedom (because there are now four classes in the χ^2 test). The observed and expected values do not deviate significantly.

To deal with parts **(b)** and **(c)** it is easier to see the observed values for the monohybrid ratios if the phenotypes are listed:

smooth, yellow	315
smooth, green	108
wrinkled, yellow	101
wrinkled, green	32

For the smooth: wrinkled *monohybrid component*, the smooth types total 423 (315 + 108), while the wrinkled types total 133 (101 + 32).

Expected ratio	Observed (o)	Expected (e)
3/4	423	417
1/4	133	139

The χ^2 value is 0.35 and in examining the K/C text for 1 degree of freedom, the p value is greater than 0.50 and less than 0.90. We fail to reject the null hypothesis and are confident that the observed values do not differ significantly from the expected values.

(c) For the yellow:green portion of the problem, see that there are 416 yellow plants (315 + 101) and 140 (108 + 32) green plants.

Expected ratio	Observed (o)	Expected (e)
3/4	416	417
1/4	140	139

The χ^2 value is 0.01 and in examining the K/C text for 1 degree of freedom, the p value is greater than 0.90. We fail to reject the null hypothesis and are confident that the observed values do not differ significantly from the expected values.

24. It would be best to set up two tables based on the two hypotheses:

Expected ratio	Observed (o)	Expected (e)
3/4	250	300
1/4	150	100

Expected ratio	Observed (o)	Expected (e)
1/2	250	200
1/2	150	200

For the test of a 3:1 ratio, the χ^2 value is 33.3 with an associated p value of less than 0.01 for 1 degree of freedom. For the test of a 1:1 ratio, the χ^2 value is 25.0 again with an associated p value of less than 0.01 for 1 degree of freedom. Based on these probability values, both null hypotheses should be rejected.

25. Use of the $p = 0.10$ as the "critical" value for rejecting or failing to reject the null hypothesis instead of $p = 0.05$ would allow more null hypotheses to be rejected. Notice in the K/C text that as the χ^2 values increase, there is a higher likelihood that the null hypothesis will be rejected because the higher values are more likely to be associated with a p value which is less than 0.05.

As the critical *p* value is increased, it takes a smaller χ^2 value to cause rejection of the null hypothesis. It would take less difference between the expected and observed values to reject the null hypothesis, therefore the stringency of failing to reject the null hypothesis is increased.

26. While there are many different inheritance patterns which will be described later in the K/C text (codominance, incomplete dominance, sex-linked inheritance, etc.) the range of solutions to this question is limited to the concepts developed in the first three chapters, namely dominance or recessiveness.

If a gene is dominant, it will not skip generations nor will it be passed to offspring unless the parents have the gene. On the other hand, genes which are recessive can skip generations and exist in a carrier state in parents. For example, notice that II-4 and II-5 produce a female child (III-4) with the affected phenotype. On these criteria alone, the gene must be viewed as being recessive. Note: if a gene is recessive and X-linked (to be discussed later) the pattern will often be from affected male to carrier female to affected male.

To provide genotypes for each individual consider that if the box or circle is shaded, the *aa* genotype is to be assigned. If offspring are affected (shaded) a recessive gene must have come from both parents.

I-1 (*Aa*), I-2 (*aa*), I-3 (*Aa*), I-4(*Aa*)

II-1 (*aa*), II-2 (*Aa*), II-3 (*aa*), II-4 (*Aa*), II-5 (*Aa*), II-6 (*aa*), II-7 (*AA* or *Aa*), II-8 (*AA* or *Aa*)

III-1 (*AA* or *Aa*), III-2 (*AA* or *Aa*), III-3 (*AA* or *Aa*), III-4 (*aa*), III-5 (probably *AA*), III-6 (*aa*)

IV-1 through IV-7 all *Aa*.

27. Applying the same logic as in question #26, the gene is inherited as an autosomal recessive. Notice that two normal individuals II-3 and II-4 have produced a daughter (III-2) with myopia.

I-1 (*aa*), I-2 (*Aa* or *AA*), I-3 (*Aa*), I-4 (*Aa*)
II-1 (*Aa*), II-2 (*Aa*), II-3 (*Aa*), II-4 (*Aa*), II-5 (*aa*), II-6 (*AA* or *Aa*), II-7 (*AA* or *Aa*)
III-1 (*AA* or *Aa*), III-2 (*aa*), III-3 (*AA* or *Aa*)

28. Given the cross *AaBbCC* X *AABbCc* we can apply the product rule which states that when two or more events occur independently but simultaneously, their combined probability is equal to the product of their individual probabilities.

The probability of getting *AA* from
   Aa   X   *AA* is 1/2
The probability of getting *Bb* from
   Bb   X   *Bb* is 1/2
The probability of getting *Cc* from
   CC   X   *Cc* is 1/2
The *overall* probability then is

   1/2   X   1/2   X   1/2 = 1/8

29. The probability of getting *aabbcc* from the *AaBbCC* X *AABbCc* mating is zero because of homozygosity for *AA* and *CC*.

30. Because all the offspring will show the dominant A and C phenotypes and 3/4 will show the B phenotype, the probability of an offspring showing all three dominant traits would be 1 X 3/4 X 1 = 3/4.

31. (a) 1/6

(b) Apply the product law: 1/6 X 1/6 = 1/36

(c) Consider that there are two ways of coming up with the 3 and 6 and it doesn't matter in which way it is achieved: $(1/6 \times 1/6) + (1/6 \times 1/6) = 1/18$. Perhaps another way to think of it is the following: the probability is $1/3$ that one die will come up a 3 or a 6, and $1/6$ that the other die will come up with the appropriate number. Therefore, the answer is again $1/3 \times 1/6 = 1/18$.

(d) $1/3$

32. Calculate each allelic pair separately and know that for each plant expressing the dominant phenotype, there is a $2/3$ probability that it is heterozygous. Since the $2/3$ probability applies to two independent loci, the product law is applied:

$$2/3 \quad \times \quad 2/3 \quad = \quad 4/9$$

33. (a) There are two possibilities. Either the trait is dominant, in which case I-1 is heterozygous as are II-2 and II-3, or the trait is recessive and I-1 is homozygous and I-2 is heterozygous. Under the condition of recessiveness, both II-1 and II-4 would be heterozygous, II-2 and II-3 homozygous.

(b) Recessive: Parents Aa, Aa

(c) Recessive: Parents Aa, Aa

(d) Recessive: Parents AA (probably), aa Second pedigree: Recessive or dominant, not sex-linked, if recessive, parents Aa, aa

(e) See initial explanation in this problem. It is identical to the first pedigree.

34. Assuming that both parents are heterozygous, the probability of being a carrier (for the male and female) is $2/3$ for each. Since a child has a $1/4$ chance of being homozygous recessive from carrier parents, the overall probability of the child having cystic fibrosis is:

$$2/3 \quad \times \quad 2/3 \quad \times \quad 1/4 = 1/9$$

35. (a) $\quad P = [n! (1/2)^5 (1/2)^0]/s!t! = 1/32$

(b) $\quad P = [5! (1/2)^3 (1/2)^2]/3!2! = 5/16$

(c) $\quad P = [5! (1/2)^2 (1/2)^3]/2!3! = 5/16$

(d) There are two ways of all being the same sex, all males or all females. Therefore the final probability is the sum of the two independent probabilities:

$$1/32 + 1/32 = 1/16.$$

36. $\quad P = [8! (3/4)^6 (1/4)^2]/6!2!$

37. (a) First consider that each parent is homozygous (true-breeding in the question) and since in the F_1 only round, axial, violet, and full phenotypes were expressed, they must each be dominant. Because all genes are on nonhomologous chromosomes, independent assortment will occur.

(b) Round, axial, violet and full would be the most frequent phenotypes:

$$3/4 \quad \times \quad 3/4 \quad \times \quad 3/4 \quad \times \quad 3/4$$

(c) Wrinkled, terminal, white, and constricted would be the least frequent phenotypes:

$$1/4 \quad \times \quad 1/4 \quad \times \quad 1/4 \quad \times \quad 1/4$$

(d)

$$1/4 \quad \times \quad 1/4 \quad \times \quad 1/4 \quad \times \quad 1/4 \quad \times \quad 2$$

(e) There would be 16 different phenotypes in the test cross offspring just as there are 16 different phenotypes in the F_2 generation.

38. (a) The first task is to draw out an accurate pedigree (one of serveral possibilities):

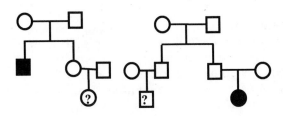

(b) The probability that the female (whose maternal uncle had TSD) is heterozygous is 1/3 because she is not TSD and her mother had a 2/3 chance of being heterozygous and she has a 1/2 chance of passing the TSD gene to her daughter (2/3 X 1/2 = 1/3). The male (whose paternal first cousin had TSD) has a 1/4 chance of being heterozygous, assuming that either (but not both, because the gene is said to be rare) his grandmother or grandfather was heterozygous. Therefore the probability that both the male and female are heterozygous is:

1/3   X   1/4 = 1/12

(c) The probability that neither is heterozygous is:

2/3   X   3/4 = 6/12

(d) The probability that one is heterozygous is:

(1/3 X 3/4)   +   (2/3 X 1/4)   = 5/12

39.

(a) Notice in cross #1 that the ratio of straight wings to curled wings is 3:1 and the ratio of short bristles to long bristles is also 3:1. This would indicate that straight is dominant to curled and short is dominant to long.

Possible symbols would be (using standard *Drosophila* symbolism):

straight wings = w^+ curled wings = w

short bristles = b^+ long bristles = b

(b)

Cross #1: w^+/w ; b^+/b	X w^+/w ; b^+/b
Cross #2: w^+/w ; b/b	X w^+/w ; b/b
Cross #3: w/w ; b/b	X w^+/w ; b^+/b
Cross #4: w^+/w^+ ; b^+/b X w^+/w^+ ; b^+/b (one parent could be w^+/w)	
Cross #5: w/w ; b^+/b	X w^+/w ; b^+/b

40. Because there are three possibilities within the group, it probably represents a case of incomplete dominance or codominance (see Chapter 4).

41. (a) First, consider that the data represent a 3:1 ratio based on the information given in the problem: *Ss* X *Ss*. Compute the expected quantities for each class by multiplying the totals by 3/4 and 1/4.

Set I Expected Numbers:

Tall = 26.25
Short = 8.75

Set II Expected Numbers:

Tall = 262.5
Short = 87.5

For Set I the χ^2 value would be:

$(30-26.25)^2/26.25$ + $(5-8.75)^2/8.75$

= 2.15 with p being between 0.2 and 0.05

so one would accept the null hypothesis of no significant difference between the expected and observed values.

For Set II, the χ^2 value would be:

21.43 and $p<0.001$ and one would reject the null hypothesis and assume a significant difference between the observed and expected values.

(b) Clearly, with an increase in sample size a different conclusion is reached. In fact, most statisticians recommend that the expected values in each class should not be less than 10. In most cases, more confidence is gained as the sample size increases, however, depending on the organism or experiment, there are practical limits on sample size.

Chapter 10: Extensions of Mendelian Genetics

Concept Areas	Corresponding Problems
Function/Symbolism of Alleles (Genes)	31
Incomplete Dominance, Codominance	1, 2, 5, 12, 13, 14, 23
Multiple Alleles	5, 6, 7, 8, 10, 11
Lethal Alleles	3, 4, 41
Gene Interaction	8, 9, 20, 22, 29, 38, 39, 41, 44
Epistasis	16, 17, 18, 19, 21, 40, 42, 46, 47
Novel Phenotypes	15
X-Linkage	24, 25, 26, 27, 28, 29, 30, 34, 44
Sex-Limited/Sex-Influenced Inheritance	32, 33, 43
Phenotpic Expression	35, 36
Genetic Anticipation and Imprinting	37

Vocabulary: Organization and Listing of Terms and Concepts

Structures and Substances

Wild type, mutations

Native antigens

Antibodies

　glycoprotein

　isoagglutinogen

Ommatidia

　drosopterin

　xanthommatin

Hexosaminidase

Heterochromatin

Lactose

Processes/Methods

Incomplete (partial) dominance

　pink flowers, 1:2:1

Codominance

　MN blood groups

Multiple allelism

　ABO blood types

　antigen-antibody reaction

　　agglutination

　H substance

　galactose

　N-acetylglucosamine

Chapter 10 Extensions of Mendelian Genetics

Bombay phenotype

  secretor locus

    Rh antigens

    erythroblastosis fetalis (HDN)

  white eye in *Drosophila*

Lethal alleles

  recessive, dominant

    yellow coat color in mice

    Huntington's disease

Gene interaction: discontinuous variation

  epigenesis

  epistasis

    hypostatic

    homozygous recessive, 9:3:4

      coat color in mice

      Bombay phenotype

    dominant

      fruit color in squash, 12:3:1

    other

      white flowers in peas, 9:7

  novel phenotypes

    fruit shape in *Cucurbita*

    eye color in *Drosophila*

  complementation

    cis, trans

X-linkage, Chromosome Theory

  hemizygous, crisscross pattern

Sex-limited inheritance

  feathering in chickens

Sex-influenced inheritance

  pattern baldness

Penetrance

Expressivity

  genetic background

  position effect

  suppression

Conditional mutation

  temperature

  nutritional mutant

    auxotroph

    phenylketonuria

    galactosemia

Tay-Sachs

Lesch-Nyhan

Duchene muscular dystrophy (DMD)

Huntington Disease

Genetic anticipation

Genomic (parental) imprinting

  Prader-Willi syndrome, Angelman syndrome

Chapter 10 Extensions of Mendelian Genetics

Concepts

Gene interaction (F10.1)

Neo-Mendelian genetics

Allele (F10.2)

    wild type, mutation

        loss of wild type function

        reduced or increased function

Symbolism (F10.2)

    recessive trait (e, e^+)

    dominant trait (Wr, Wr^+)

    no dominance ($R1$, $R2$) (L^M, L^N)

    leu-, leu+, dnaA, BRCA1

Modified ratios (T10.1), lethality

    3:1, 1:2:1, 9:3:3:1

      3:6:3:1:2:1

        9:3:4, 12:3:1, 9:7, 1:4:6:4:1

Complementation

X-linked inheritance

Sex-limited inheritance

Sex-influenced inheritance

Phenotypic expression

    penetrance, expressivity

Genetic background

Environmental effects

T10.1 Examples of typical monohybrid and dihybrid ratios with several modifications.

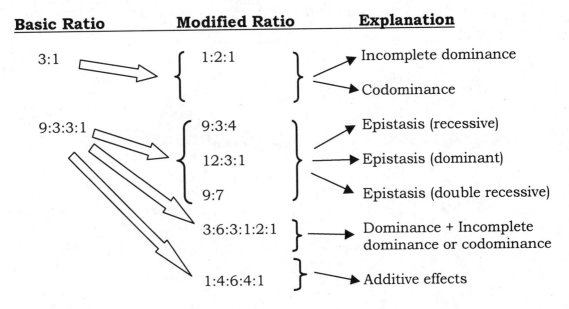

Basic Ratio	Modified Ratio	Explanation
3:1	1:2:1	Incomplete dominance
		Codominance
9:3:3:1	9:3:4	Epistasis (recessive)
	12:3:1	Epistasis (dominant)
	9:7	Epistasis (double recessive)
	3:6:3:1:2:1	Dominance + Incomplete dominance or codominance
	1:4:6:4:1	Additive effects

F10.1 Illustration of gene interaction where products from more than one gene pair influence one characteristic or phenotypic trait.

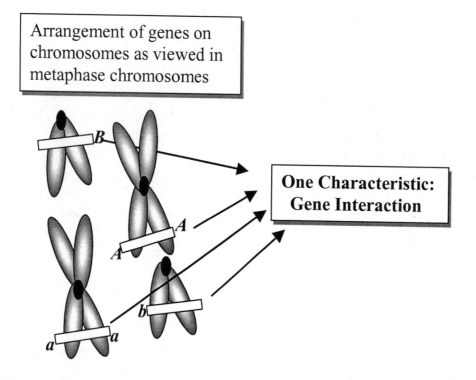

Arrangement of genes on chromosomes as viewed in metaphase chromosomes

One Characteristic: Gene Interaction

Example: Two gene pairs influencing the pigmentation pattern on the shark. Various gene products contribute in a variety of ways to generate a particular pigment pattern.

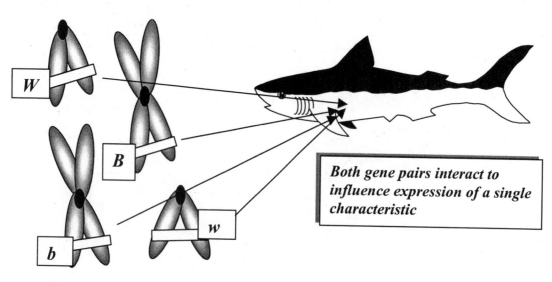

Both gene pairs interact to influence expression of a single characteristic

F10.2 Symbolism associated with the wild type activity of a gene and several possible outcomes of the mutant state: **A.** wild type; **B.** too much product; **C.** too little product; **D.** no product; **E.** both products expressed; **F.** reduced product.

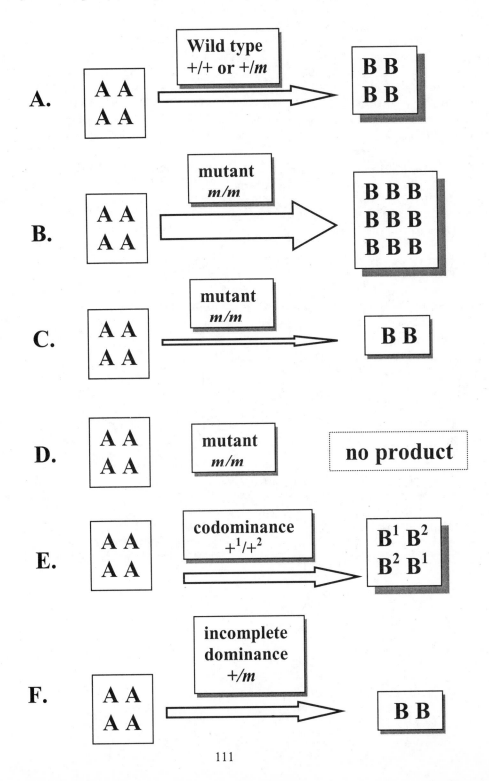

Chapter 10 Extensions of Mendelian Genetics

Solutions to Problems and Discussion Questions

1. In the first sentence of this problem, notice that there is one characteristic (coat color) and three phenotypes mentioned; red, white, or roan. The fact that roan is intermediate between red and white suggests that this may be a case of incomplete dominance, with roan being the intermediate and therefore the heterozygous type. If that is the case then we should suspect a 1:2:1 phenotypic ratio in crosses of "roan to roan."

Looking at the data given, notice that a cross of the "extremes" (red X white) gives roan, suggesting its heterozygous nature and the homozygous nature of the parents. Seeing the 1:2:1 ratio in the offspring of

roan X roan

confirms the hypothesis of incomplete dominance as the mode of inheritance.

Symbolism:
AA = red,
aa = white,
Aa = roan

Crosses: It is important at this point that you not be fully dependent on writing out complete Punnett squares for each cross. Begin working these simple problems in your head.

AA	X	*AA*	⟹	*AA*
aa	X	*aa*	⟹	*aa*
AA	X	*aa*	⟹	*Aa*
Aa	X	*Aa*	⤵	
1/4 *AA*; 2/4 *Aa*; 1/4 *aa*				

2. *Incomplete dominance* can be viewed more as a quantitative phenomenon where the heterozygote is intermediate (approximately) between the limits set by the homozygotes. Pink is intermediate between red and white.

Codominance can be viewed in a more qualitative manner where both of the alleles in the heterozygote are expressed. For example in the AB blood group, both the I^A and I^B genes are expressed. There is no intermediate class which is part I^A and I^B.

3. Notice that there is one typical (coat color) and one atypical (lethality) characteristic mentioned. Often under this condition of two characteristics, we must decide if the problem involves one or more than one gene pair. Because the genotypes are given here it is obvious that lethality is associated with expression of the coat color alleles and therefore one gene pair is involved. This is a monohybrid condition.

Pp X *Pp* ⤵
1/4 *PP* (**lethal**)
2/4 *Pp* (platinum)
1/4 *pp* (silver)

Therefore the ratio of surviving foxes is 2/3 platinum, 1/3 silver. The *P* allele behaves as a recessive in terms of lethality (seen only in the homozygote) but as a dominant in terms of coat color (seen in the homozygote).

4. In this problem it would be helpful to first diagram the phenotypes of the crosses so that you can get some idea of the inheritance pattern. From those phenotypic crosses, a suggestion as to the genotypes can be made and verified.

> Cross 1:
>
> short tail   X   normal long tail ⤵
>
> approximately   1/2 short, 1/2 long

This tells you that one type is heterozygous, the other homozygous.

> Cross 2:
>
> short tail   X   short tail ⤵
>
> 6 short tail, 3 long tail (2/3 short, 1/3 long)

At this point one would consider that the 2/3 *short* are heterozygotes, and *long* is the homozygous class. Also short is dominant to long. Since these ratios were repeated and verified, one can conclude that a 2:1 ratio is not a statistical artifact and that the following genotypic model would hold. Because long is the "normal" does not mean that it is dominant.

> Symbolism:
>
> *S* = short, *s* = long

> Cross 1:
>
> *Ss* X *ss* ⤵
>
> 1/2 *Ss* (short), 1/2 *ss* (long)

> Cross 2:
>
> *Ss*   X   *Ss* ⤵
>
> 1/4 *SS* (lethal), 2/4 *Ss* (short), 1/4 *ss* (long)

5. In the section on multiple alleles in the K/C text, there is a table which indicates all the genotypes requested.

Blood Group (phenotype)	Genotype(s)
A	$I^A I^A, I^A I^o$
B	$I^B I^B, I^B I^o$
AB	$I^A I^B$
O	$I^o I^o$

I^A and I^B are codominant (notice the AB blood group) while being dominant to I^o.

6. In this problem remember that individuals with blood type B can have the genotype $I^B I^B$ or $I^B I^o$ and those with blood type A, genotypes $I^A I^A$ or $I^A I^o$.

Male Parent: must be $I^B I^o$ because the mother is $I^o I^o$ and one inherits one homologue (therefore one allele) from each parent.

Female Parent: must be $I^A I^o$ because the father is $I^B I^o$ and one inherits one homologue (therefore one allele) from each parent. The father can not be $I^B I^B$ and have a daughter of blood type A.

Offspring:

$I^A I^o$ X $I^B I^o$

	I^B	I^o
I^A	$I^A I^B$(AB)	$I^A I^o$(A)
I^o	$I^B I^o$ (B)	$I^o I^o$(O)

The ratio would be

1(A):1(B):1(AB):1(O).

7. Given that a child is blood type O (genotype I^oI^o) and the mother blood type A (she must be I^AI^o to have had a type O child), then the father could have the following genotypes: I^BI^o, I^AI^o or I^oI^o. In other words, the father must have been able to contribute I^o to the child.

The only *blood type* which would exclude a male from being the father would be AB, because no I^o allele is present. Because many individuals in a population could have genotypes with the I^o allele, one could not prove that a particular male was the father by this method.

8.

Symbolism:

Se = secretor se = nonsecretor

$I^A I^B$ $Sese$ X $I^O I^O$ $Sese$

(a)

	$I^O Se$	$I^O se$
$I^A Se$	A, secretor	A, secretor
$I^A se$	A, secretor	A, nonsecretor
$I^B Se$	B, secretor	B, secretor
$I^B se$	B, secretor	B, nonsecretor

Overall ratio: 3/8 A, secretor
1/8 A, nonsecretor
3/8 B, secretor
1/8 B, nonsecretor

(b) 1/4 of all individuals will have blood type O.

9. Consider the term "gene interaction" at two levels: *allelic* and *non-allelic*. Allelic interactions include *dominance, recessiveness, incomplete dominance,* and *codominance*. Non-allelic gene interactions are more the focus of this chapter, where one or two genes of a pair interact with other genes (or gene pairs) to influence the phenotype. Various forms of such interaction include *novel phenotypes, epistasis,* and *additive alleles*.

10. It is important to see that this problem involves multiple alleles, meaning that monohybrid type ratios are expected, and that there is an order of dominance which will allow certain alleles to be "hidden" in various heterozygotes. As with most genetics problems, one must look at the phenotypes of the offspring to assess the genotypes of the parents.

(a)

> Phenotypes:
>
> Himalayan   X   Himalayan ⟹ albino
>
> Genotypes: $c^h c^a$ $c^h c^a$ $c^a c^a$
>
> The Himalayan parents must both be heterozygous to produce an albino offspring.

> Phenotypes:
>
> full color   X   albino ⟹ chinchilla
>
> Genotypes: Cc^{ch} $c^a c^a$ $c^{ch} c^a$

Because of the cc albino parent, the genotype of the chinchilla F_1 must be $c^{ch}c^a$. Also in order to have a chinchilla offspring at all, the full color parent must he heterozygous for chinchilla.

Therefore the cross of albino with chinchilla would be as follows:

> $c^a c^a$ X $c^{ch} c^a$ ⟹
>
> 1/2 chinchilla; 1/2 albino

(b)

Phenotypes:

   albino   X   chinchilla ⟹ albino

Genotypes: $c^a c^a$ $c^{ch} c^a$ $c^a c^a$

Phenotypes:

   full color   X   albino ⟹ full color

Genotypes: $C\_$ $c^a c^a$ $C c^a$

It is impossible to determine the complete genotype of the full color parent but the full color offspring must be as indicated, $C c^a$.

Therefore the cross of the albino with full color would be as follows:

$c^a c^a$ X $C c^a$ ⤵

1/2 full color; 1/2 albino

(c)

Phenotypes:

chinchilla   X   albino ⟹ Himalayan

Genotypes: $c^{ch} c^h$ $c^a c^a$ $c^h c^a$

The chinchilla parent must be heterozygous for Himalayan because of the Himalayan offspring.

Phenotypes:

   full color   X   albino ⟹ Himalayan

Genotypes: $C c^h$ $c^a c^a$ $c^h c^a$

Therefore a cross between the two Himalayan types would produce the following offspring:

$c^h c^a$ X $c^h c^a$
⇩
3/4 Himalayan; 1/4 albino

11. It is important to see that this problem involves multiple alleles, meaning that monohybrid type ratios are expected, and that there is an order of dominance which will allow certain alleles to be "hidden" in various heterozygotes. As with most genetics problems, one must look at the phenotypes of the offspring to assess the genotypes of the parents.

(a)

Parents: sepia   X   cream

Because both guinea pigs had albino parents, both are heterozygous for the ca allele.

Cross:

$c^k c^a$ X $c^d c^a$

2/4 sepia; 1/4 cream; 1/4 albino

(b)

Parents: sepia   X   cream

Because the sepia parent had an albino parent it must be $c^k c^a$. Because the cream guinea pig had two sepia parents

($c^k c^d$ X $c^k c^d$ or $c^k c^d$ X $c^k c^a$),

the cream parent could be $c^d c^d$ or $c^d c^a$.

Crosses: $c^k c^a$ X $c^d c^d$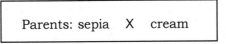

1/2 sepia; 1/2 cream*
*(if parents are assumed to be homozygous)

or $c^k c^a$ X $c^d c^a$ ⤵

1/2 sepia; 1/4 cream; 1/4 albino

(c)

> Parents: sepia X cream

Because the sepia guinea pig had two full color parents which could be

$$Cc^k, \quad Cc^d, \quad \text{or} \quad Cc^a$$

(not CC because sepia could not be produced), its genotype could be

$$c^k c^k, \quad c^k c^d, \quad \text{or} \quad c^k c^a.$$

Because the cream guinea pig had two sepia parents

$$(c^k c^d \quad X \quad c^k c^d \quad \text{or} \quad c^k c^d \quad X \quad c^k c^a),$$

the cream parent could be $c^d c^d$ or $c^d c^a$.

> **Crosses:**
>
> $c^k c^k$ X $c^d c^d$ ⟹ all sepia
>
> $c^k c^k$ X $c^d c^a$ ⟹ all sepia
>
> $c^k c^d$ X $c^d c^d$ ⟹ 1/2 sepia; 1/2 cream
>
> $c^k c^d$ X $c^d c^a$ ⟹ 1/2 sepia; 1/2 cream
>
> $c^k c^a$ X $c^d c^d$ ⟹ 1/2 sepia; 1/2 cream
>
> $c^k c^a$ X $c^d c^a$ ⟹
>
> 1/2 sepia; 1/4 cream; 1/4 albino

(d)

> Parents: sepia X cream

Because the sepia parent had a full color parent and an albino parent (Cc^k X $c^a c^a$), it must be $c^k c^a$. The cream parent had two full color parents which could be Cc^d or Cc^a; therefore it could be $c^d c^d$ or $c^d c^a$.

> **Crosses:**
>
> $c^k c^a$ X $c^d c^d$ ⟹ 1/2 sepia; 1/2 cream
>
> $c^k c^a$ X $c^d c^a$ ⟹
>
> 1/2 sepia; 1/4 cream; 1/4 albino

12. Three independently assorting characteristics are being dealt with: flower color (incomplete dominance), flower shape (dominant/recessive), and plant height (dominant/recessive). Establish appropriate gene symbols:

> **Flower color:**
>
> RR = red ; Rr = pink; rr = white
>
> **Flower shape:**
>
> P = personate; p = peloric
>
> **Plant height:**
>
> D = tall; d = dwarf

(a)

> $RRPPDD$ X $rrppdd$ ⟹
>
> $RrPpDd$ (pink, personate, tall)

(b) Use *components* of the forked line method as follows:

> 2/4 pink X 3/4 personate X 3/4 tall
>
> = 18/64

13. There are two characteristics, flower color and flower shape. Because pink results from a cross of red and white, one would conclude that flower color is "monohybrid" with incomplete dominance.

Chapter 10 Extensions of Mendelian Genetics

In addition, because personate is seen in the F_1 when personate and peloric are crossed, personate must be dominant to peloric. Results from crosses (c) and (d) verify these conclusions. The appropriate symbols would be as follows

---

Flower color:

RR = red; Rr = pink; rr = white

Flower shape:

P = personate; p = peloric

(a)

$RRpp$ X $rrPP \Longrightarrow RrPp$

(b)

$RRPP$ X $rrpp \Longrightarrow RrPp$

(c)

$RrPp$ X $RRpp \Longrightarrow$ $\begin{cases} RRPp \\ RRpp \\ RrPp \\ Rrpp \end{cases}$

(d)

$RrPp$ X $rrpp \Longrightarrow$ $\begin{cases} rrPp \\ rrpp \\ RrPp \\ Rrpp \end{cases}$

In the cross of the F_1 of (a) to the F_1 of (b), both of which are double heterozygotes, one would expect the following:

$RrPp$ X $RrPp$

1/4 red ⟨ 3/4 personate ---> 3/16 red, personate
1/4 peloric ---> 1/16 red, peloric

2/4 pink ⟨ 3/4 personate ---> 6/16 pink, personate
1/4 peloric ---> 2/16 pink, peloric

1/4 white ⟨ 3/4 personate ---> 3/16 white, personate
1/4 peloric ---> 1/16 white, peloric

14. (a) This is a case of incomplete dominance in which, as shown in the third cross, the heterozygote (palomino) produces a typical 1:2:1 ratio. Therefore one can set the following symbols:

$C^{ch}C^{ch}$ = chestnut

C^cC^c = cremello

$C^{ch}C^c$ = palomino

(b) The F_1 resulting from matings between cremello and chestnut horses would be expected to be all palomino. The F_2 would be expected to fall in a 1:2:1 ratio as in the third cross in part (a) above.

15. This is a case of gene interaction (novel phenotypes) in which the recessive, independently assorting genes *brown* and *scarlet* (both recessive) interact to give the white phenotype. Refer to the K/C text and see that the symbolism uses a "+" superscript to indicate the wild type. For simplicity in this problem, assume that all parental crosses involve homozygotes.

(a) $bw^+/bw^+;st^+/st^+$ X $bw/bw;st/st$

$\Downarrow$

$bw^+/bw;\ st^+/st$ (wild)

The F$_2$ would produce the expected 9:3:3:1 ratio except that gene interaction will give the white phenotype in the 1/16 class.

$bw^+/\_ ; st^+/\_$	= wild type
$bw^+/\_; st/st$	= scarlet
$bw/bw ; st^+/\_$	= brown
$bw/bw ; st/st$	= white

(b)

$bw^+/bw^+;st^+/st^+$ X $bw^+/bw^+;st/st$
⇓
$bw^+/bw^+; st^+/st$ (wild)

The F$_2$, resulting from a cross of

$$bw^+/bw^+; st^+/st \quad X \quad bw^+/bw^+; st^+/st$$

would produce a 3:1 ratio of wild to scarlet.

(c)

$bw/bw;st^+/st^+$ X $bw/bw;st/st$
⇓
$bw/bw; st^+/st$ (brown)

The F$_2$, resulting from a cross of

$$bw/bw; st^+/st \quad X \quad bw/bw; st^+/st$$

would produce a 3:1 ratio of brown to white. Notice in the F$_2$ crosses in parts (b) and (c) one of the parents in each is homozygous, therefore the crosses will give monohybrid types of ratios.

16. This is a case in which epistasis (from *cc*) results in a "masking" of genes at the *A* locus. In this case there will be modifications of typical 9:3:3:1 and 1:1:1:1 ratios because of gene interactions.

(a) In a cross of

$$AACC \quad X \quad aacc,$$

the offpsring are all *AaCc* (agouti) because the *C* allele allows pigment to be deposited in the hair and when it is it will be agouti. F$_2$ offspring would have the following "simplified" genotypes with the corresponding phenotypes:

$A\_C\_$ = 9/16 (agouti)

A_cc = 3/16
(colorless because *cc* is epistatic to *A*)

$aaC\_$ = 3/16 (black)

$aacc$ = 1/16
(colorless because *cc* is epistatic to *aa*)

The two colorless classes are phenotyically indistinguishable, therefore the final ratio is 9:3:4.

(b) Results of crosses of female agouti

$$(A\_C\_) \quad X \quad aacc \text{ (males)}$$

are given in three groups:

(1) To produce an even number of agouti and colorless offspring, the female parent must have been *AACc* so that half of the offspring are able to deposit pigment because of *C* and when they do, they are all agouti(having received only *A* from the female parent).

(2) To produce an even number of agouti and black offspring the mother must have been *Aa* and so that no colorless offspring were produced, the female must have been *CC*. Her genotype must have been *AaCC*.

(3) Notice that half of the offspring are colorless, therefore the female must have been *Cc*. Half of the pigmented offspring are black and half are agouti, therefore the female must have been *Aa*. Overall, the *AaCc* genotype seems appropriate.

Chapter 10 *Extensions of Mendelian Genetics*

17. Notice that the distribution of observed offspring fits a 9:3:4 ratio quite well. This suggests that two independently assorting gene pairs with epistasis are involved. Assign gene symbols in the usual manner:

> A = pigment; a = pigmentless (colorless)
>
> B = purple; b = red

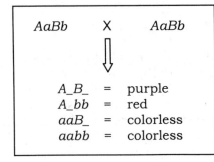

One may see this occurring in the following manner:

precursor ---+-> cyanidin ---+-> purple pigment

(colorless) *aa*   (red) *bb*

18. This is a case of gene interaction (novel phenotypes) where the yellow and black types (double mutants) interact to give the cream phenotype and epistasis where the *cc* genotype produces albino.

(a)

AaBbCc ⟹ gray (*C* allows pigment)

(b)

A_B_Cc ⟹ gray (*C* allows pigment)

(c) Use the forked line method for this portion

3/4 A_
— 3/4 B_ — 1/2 Cc ⟹ 9/32 gray
— 1/2 cc ⟹ 9/32 albino
— 1/4 bb — 1/2 Cc ⟹ 3/32 yellow
— 1/2 cc ⟹ 3/32 albino

1/4 aa
— 3/4 B_ — 1/2 Cc ⟹ 3/32 black
— 1/2 cc ⟹ 3/32 albino
— 1/4 bb — 1/2 Cc ⟹ 1/32 cream
— 1/2 cc ⟹ 1/32 albino

Combining the phenotypes gives (always count the proportions to see that they add up to 1.0):

16/32 albino;

9/32 gray;

3/32 yellow;

3/32 black;

1/32 cream

(d) Use the forked line method for this portion

3/4 A_ ⟹ 1/1 BB — 3/4 C_ ⟹ 9/16 (gray)
— 1/4 cc ⟹ 3/16 (albino)

1/4 aa ⟹ 1/1 BB — 3/4 C_ ⟹ 3/16 (black)
— 1/4 cc ⟹ 1/16 (albino)

Combining the phenotypes gives (always count the proportions to see that they add up to 1.0):

9/16 (gray);

3/16 (black);

4/16 (albino)

119

(e) Use the forked line method for this portion

$$1/1\ AA \begin{cases} 3/4\ B\_ \begin{cases} 1/2\ Cc \Rightarrow 3/8\ \text{(gray)} \\ 1/2\ cc \Rightarrow 3/8\ \text{(albino)} \end{cases} \\ 1/4\ bb \begin{cases} 1/2\ Cc \Rightarrow 1/8\ \text{(yellow)} \\ 1/2\ cc \Rightarrow 1/8\ \text{(albino)} \end{cases} \end{cases}$$

The final ratio would be

3/8 (grey);

1/8 (yellow);

4/8 (albino)

19. Treat each of the crosses as a series of monohybrid crosses, remembering that albino is epistatic to color and black and yellow interact to give cream.

(a) Since this is a 9:3:3:1 ratio with no albino phenotypes, the parents must each have been double heterozygotes and incapable of producing the *cc* genotype.

Genotypes:
AaBbCC X *AaBbCC*
or
AaBbCC X *AaBbCc*
Phenotypes:
gray X gray.

(b) Since there are no black offspring, there are no combinations in the parents which can produce *aa*. The 4/16 proportion indicates that the *C* locus is heterozygous in both parents.

If the parents are

AABbCc X AaBbCc
or
AABbCc X AABbCc

then the results would follow the pattern given.

Phenotypes: gray X gray.

(c) Notice that 16/64 or 1/4 of the offspring are albino, therefore the parents are both heterozygous at the *C* locus. Second, notice that without considering the *C* locus, there is a 27:9:9:3 ratio which reduces to a 9:3:3:1 ratio. Given this information, the genotypes must be

AaBbCc X *AaBbCc*.
Phenotypes: gray X gray

(d) Notice that 2/8 or 1/4 of the offspring are albino which indicates that both parents are *Cc*. Also notice that the ratio of black to cream is 1:1 suggesting that the parents are *Bb* and *bb*. Because there are no grey or yellow offspring, there can be no *A* alleles.

Genotypes:
aaBbCc X *aabbCc*
Phenotypes: black X cream.

(e) Notice that half of the offspring are albino indicating that, for the *C* locus the genotypes are *Cc* and *cc*. There is a 3:1 ratio of black to cream indicating heterozygosity for the *B* locus in each parent. Because there are no gray or yellow offspring, there can be no *A* alleles.

Genotypes:
aaBbCc X *aaBbcc*
Phenotypes: black X albino

20. After reading the problem, glance at the kinds of F_1 and F_2 ratios. Notice that the first two, (A) and (B), appear as monohybrid ratios and (C) is clearly dihybrid. You need to see this combination of crosses as being solved with one set of gene symbols. The fact that cross (c) yields a 9:3:3:1 ratio gives you a start.

(a) Going back to basics, set up the relationship which you know holds for a 9:3:3:1 ratio as follows:

A_B_ = 9/16

A_bb = 3/16

aaB_ = 3/16

aabb = 1/16

Then assign the phenotypes from cross (C) as indicated:

A_B_ = 9/16 (green)

A_bb = 3/16 (brown)

aaB_ = 3/16 (grey)

aabb = 1/16 (blue)

Now it should become clear that blue results from interaction of the *aa* and *bb* genotypes and brown and grey result from homozygosity of either of the two genes as shown above. From this model, see if crosses (a), (b), and (c) and the resulting progeny make sense.

Cross A:

P$_1$:
  AABB X aaBB

F$_1$: AaBB

F$_2$: 3/4 A_BB: 1/4 aaBB

Cross B:

P$_1$: AABB X AAbb

F$_1$: AABb

F$_2$: 3/4 AAB_ : 1/4 AAbb

Cross C:

P$_1$: aaBB X AAbb

F$_1$: AaBb

F$_2$: 9/16 A_B_ : 3/16 A_bb: 3/16 aaB_ : 1/16 aabb

(b) This question is exactly as that in Cross C. The genotype of the unknown P$_1$ individual would be *AAbb* (brown) while the F$_1$ would be *AaBb* (green).

21. First see in this problem that a 9:7 ratio is involved which implies a dihybrid condition with epistasis. Going back to a basic 9:3:3:1 ratio one can see that if the 3:3:1 groups were lumped together the 9:7 ratio would result. Assign tall to any plant with both *A* and *B*, and any dwarf plant which is homozygous for either or both the recessive alleles. The initial cross must have been

AABB X *aabb.*

There are two gene pairs involved.

(a)
A_B_ = 9/16 (tall)

A_bb = 3/16 (dwarf)

aaB_ = 3/16 (dwarf)

aabb = 1/16 (dwarf)

(b) There are three different classes of dwarf plants. Within each of the 3/16 classes there are two types:

$$A\_bb = 3/16 \text{ (dwarf)}$$

$$= 1/3 \text{ } AAbb \text{ and } 2/3 \text{ } Aabb$$

and

$$aaB\_ = 3/16 \text{ (dwarf)}$$

$$= 1/3 \text{ } aaBB \text{ and } 2/3 \text{ } aaBb$$

Therefore the true breeding dwarf plants would be the following:

$$AAbb, \text{ } aaBB, \text{ and } aabb$$

and they would constitute 3/7 of the dwarf group.

22. Problems of this type often pose difficulties for students. It is important for students to go back to basic patterns of inheritance when getting started. First, see that a 9:3:3:1 ratio is involved as indicated below:

$$A\_B\_ = 9/16$$

$$A\_bb = 3/16$$

$$aaB\_ = 3/16$$

$$aabb = 1/16$$

(a) Assign the phenotypes as given, then see if patterns emerge.

$$A\_B\_ = 9/16 \text{ (yellow)}$$

$$A\_bb = 3/16 \text{ (blue)}$$

$$aaB\_ = 3/16 \text{ (red)}$$

$$aabb = 1/16 \text{ (mauve)}$$

From this information, the genotypes for the various phenotypes and the solution to the problem become clear. As stated in the problem all colors *may* be true-breeding. See that each type can exist as a full homozygote. If plants with blue flowers (homozygotes) are crossed to red-flowered homozygotes, the F_1 plants would have yellow flowers. Also as stated in the problem, if yellow-flowered plants are crossed with mauve-flowered plants, the F_1 plants are yellow and the F_2 will occur in a 9:3:3:1 ratio. All of the observations fit the model as proposed.

(b) If one crosses a true-breeding red plant ($aaBB$) with a mauve plant ($aabb$), the F_1 should be red ($aaBb$). The F_2 would be as follows:

> $aaBb$ X $aaBb$ ⟶
>
> 3/4 $aaB\_$ (red): 1/4 $aabb$ (mauve)

23. First, make certain that you understand the genetics of all the gene pairs being described in the problem. The ABO system involves multiple alleles, codominance, and dominance. The MN system is codominant while the Rh system as presented here is a dominant/recessive system. The easiest way to approach these types of problems is to consider those gene pairs which produce a low number of options in the offspring. Notice in cross #1 there are two options in the offspring for the ABO system (types A and O), but only one option for the MN system (type MN) and one option for the Rh system (Rh⁻). By looking at the most restrictive classes (MN and Rh systems in this case) one can see that option (c) is the only one which is both MN and Rh⁻. The remainder of the combinations can be determined using the same logic.

Cross #1 = (c)
Cross #2 = (d)
Cross #3 = (b)
Cross #4 = (e)
Cross #5 = (a)

There are no other combinations of solutions.

24. In order to solve this problem one must first see the possible genotypes of the parents and the grandfathers. Since the gene is X-linked the cross will be symbolized with the X chromosomes.

RG = normal vision; rg = color-blind

Mother's father: X^{rg}/Y

Father's father: X^{rg}/Y

Mother: $X^{RG}X^{rg}$

Father: X^{RG}/Y

Notice that the mother must be heterozygous for the rg allele (being normal-visioned and having inherited an X^{rg} from her father) and the father, because he has normal vision, must be X^{RG}. The fact that the father's father is color-blind does not mean that the father will be color-blind. On the contrary, the father will inherit his X chromosome from his mother.

$$X^{RG} X^{rg} \quad X \quad X^{RG}/Y$$

$$
\begin{array}{ll}
X^{RG}X^{RG} & = 1/4 \text{ daughter normal} \\
X^{RG}X^{rg} & = 1/4 \text{ daughter normal} \\
X^{RG}/Y & = 1/4 \text{ son normal} \\
X^{rg}/Y & = 1/4 \text{ son color-blind}
\end{array}
$$

Looking at the distribution of offspring:

(a) 1/4

(b) 1/2

(c) 1/4

(d) zero

25. The mating is $\quad X^{RG}X^{rg}; I^A I^O \quad X \quad X^{RG}Y; I^A I^O$

Based on the son who is colorblind and blood type O, the mother must have been heterozygous for the RG locus and both parents must have had one copy of the I^O gene. The probability of having a female child is 1/2, that she has normal vision is 1 (because the father's X is normal) and 1/4 type O blood. The final product of the independent probabilities is

$$1/2 \quad X \quad 1 \quad X \quad 1/4 \quad = \quad 1/8$$

26. Symbolism: Normal wing margins = sd^+; scalloped = sd

(a)

P_1: $X^{sd}X^{sd} \quad X \quad X^+/Y$

F_1: 1/2 X^+X^{sd} (female, normal)

1/2 X^{sd}/Y (male, scalloped)

F_2: 1/4 X^+X^{sd} (female, normal)

1/4 $X^{sd}X^{sd}$ (female, scalloped)

1/4 X^+/Y (male, normal)

1/4 X^{sd}/Y (male, scalloped)

(b)

P_1: $X^+/X^+ \quad X \quad X^{sd}/Y$

F_1: 1/2 X^+X^{sd} (female, normal)

1/2 X^+/Y (male, normal)

F_2: 1/4 X^+X^+ (female, normal)

1/4 X^+X^{sd} (female, normal)

1/4 X^+/Y (male, normal)

1/4 X^{sd}/Y (male, scalloped)

If the *scalloped* gene were not X-linked, then all of the F_1 offspring would be wild (phenotypically) and a 3:1 ratio of normal to scalloped would occur in the F_2.

27. Assuming that the parents are homozygous, the crosses would be as follows. Notice that the X symbol may remain to remind us that the *sd* gene is on the X chromosome. It is extremely important that one account for both the mutant genes and each of their wild type alleles.

P₁: $X^{sd}X^{sd}$; e^+/e^+ X X^+/Y; e/e

F₁:

   1/2 X^+X^{sd}; e^+/e (female, normal)

   1/2 X^{sd}/Y; e^+/e (male, scalloped)

F₂:

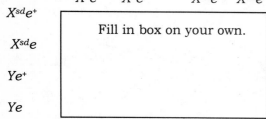

	X^+e^+	X^+e	$X^{sd}e^+$	$X^{sd}e$
$X^{sd}e^+$				
$X^{sd}e$	Fill in box on your own.			
Ye^+				
Ye				

Phenotypes:

   3/16 normal females

   3/16 normal males

   1/16 ebony females

   1/16 ebony males

   3/16 scalloped females

   3/16 scalloped males

   1/16 scalloped, ebony females

   1/16 scalloped, ebony males

Forked-line method:

P₁: $X^{sd}X^{sd}$; e^+/e^+ X X^+/Y; e/e

F₁: 1/2 X^+X^{sd}; e^+/e (female, normal)
   1/2 X^{sd}/Y; e^+/e (male, scalloped)

F₂:

	Wings	Color	
1/4	females, normal	3/4 normal	3/16
		1/4 ebony	1/16
1/4	females, scalloped	3/4 normal	3/16
		1/4 ebony	1/16
1/4	males, normal	3/4 normal	3/16
		1/4 ebony	1/16
1/4	males, scalloped	3/4 normal	3/16
		1/4 ebony	1/16

28. Set up the symbolism and the cross in the following manner:

P₁: X^+X^+; *su-v/su-v* X X^v/Y; *su-v⁺/su-v⁺*

   F₁: 1/2 X^+X^v; *su-v⁺/su-v* (female, normal)
   1/2 X^+/Y; *su-v⁺/su-v* (male, normal)

F₂: 2/4 females, 3/4 *su-v⁺/_*
     $X^+/_$ 1/4 *su-v/su-v*

   1/4 males, 3/4 *su-v⁺/_*
     X^+/Y 1/4 *su-v/su-v*

   1/4 males, 3/4 *su-v⁺/_*
     X^v/Y 1/4 *su-v/su-v*

8/16 wild type females; (none of the females are homozygous for the *vermilion* gene)

5/16 wild type males; (4/16 because they have no *vermilion* gene and 1/16 because the X-linked, hemizygous *vermilion* gene is suppressed by *su-v/su-v*)

3/16 vermilion males; (no suppression of the *vermilion* gene)

29. It is extremely important that one account for both the mutant genes and each of their wild type alleles.

(a) P$_1$: X^vX^v; $+/+$ X X^+/Y; b^r/b^r

F$_1$:

1/2 X^+X^v; $+/b^r$ (female, normal)
1/2 X^v/Y; $+/b^r$ (male, vermilion)

F$_2$:

Eye color (X)   Eye color (autosomal)

1/4 females, < 3/4 normal 3/16
   normal   1/4 brown   1/16

1/4 females, < 3/4 normal 3/16
   vermilion 1/4 brown   1/16

1/4 males, < 3/4 normal 3/16
   normal   1/4 brown   1/16

1/4 males, < 3/4 normal 3/16
   vermilion 1/4 brown   1/16

3/16 = females, normal
1/16 = females, brown eyes
3/16 = females, vermilion eyes
1/16 = females, white eyes
3/16 = males, normal
1/16 = males, brown eyes
3/16 = males, vermilion eyes
1/16 = males, white eyes

(b)

P$_1$: X^+X^+; b^r/b^r X X^v/Y; $+/+$

F$_1$:

1/2 X^+X^v; $+/b^r$ (female, normal)

1/2 X^+/Y; $+/b^r$ (male, normal)

F$_2$:

Eye color (X)   Eye color(autosomal)

2/4 females, < 3/4 normal
   normal   1/4 brown

1/4 males, < 3/4 normal
   normal   1/4 brown

1/4 males, < 3/4 normal
   vermilion 1/4 brown

6/16 = females, normal
2/16 = females, brown eyes
3/16 = males, normal
1/16 = males, brown eyes
3/16 = males, vermilion eyes
1/16 = males, white eyes

(c)

P$_1$: X^vX^v; b^r/b^r X X^+/Y; $+/+$

F$_1$: 1/2 X^+X^v; $+/b^r$ (female, normal)
      1/2 X^v/Y; $+/b^r$ (male, vermilion)

F$_2$:

Eye color (X)   Eye color (autosomal)

1/4 females, < 3/4 normal
   normal   1/4 brown

1/4 females, < 3/4 normal
   vermilion 1/4 brown

1/4 males, < 3/4 normal
   normal   1/4 brown

1/4 males, < 3/4 normal
   vermilion 1/4 brown

3/16 = females, normal
1/16 = females, brown eyes
3/16 = females, vermilion eyes
1/16 = females, white eyes
3/16 = males, normal
1/16 = males, brown eyes
3/16 = males, vermilion eyes
1/16 = males, white eyes

30.

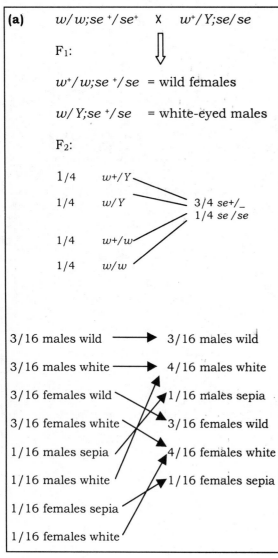

(a) $w/w;se^+/se^+$ X $w^+/Y;se/se$

F₁:

$w^+/w;se^+/se$ = wild females

$w/Y;se^+/se$ = white-eyed males

F₂:

1/4 $w+/Y$
1/4 w/Y 3/4 $se+/\_$
                     1/4 se/se
1/4 $w+/w$
1/4 w/w

3/16 males wild ⟶ 3/16 males wild

3/16 males white ⟶ 4/16 males white

3/16 females wild   1/16 males sepia

3/16 females white   3/16 females wild

1/16 males sepia   4/16 females white

1/16 males white   1/16 females sepia

1/16 females sepia

1/16 females white

(b)

$w^+/w^+;se/se$ X $w/Y;se^+/se^+$

F₁: $w^+/w;se^+/se$ = wild females
       $w^+/Y;se^+/se$ = wild males

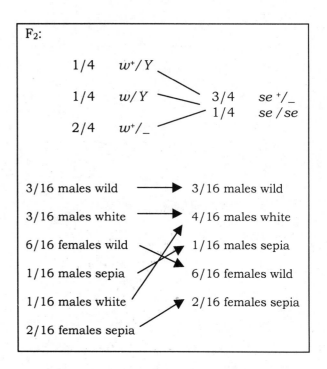

F₂:

1/4 w^+/Y
1/4 w/Y 3/4 $se^+/\_$
                      1/4 se/se
2/4 $w^+/\_$

3/16 males wild ⟶ 3/16 males wild

3/16 males white ⟶ 4/16 males white

6/16 females wild   1/16 males sepia

1/16 males sepia   6/16 females wild

1/16 males white   2/16 females sepia

2/16 females sepia

31. (a,b) In looking at the pedigrees, one can see that the condition can not be dominant because it appears in the offspring (II-3 and II-4) and not the parents in the first two cases. The condition is therefore *recessive*. In the second cross, note that the father is not shaded yet the daughter (II-4) is. If the condition is recessive, then it must also be *autosomal*.

(c)   II-1   = AA or Aa

II-6   = AA or Aa

II-9   = Aa

32. In seeing that the distribution of phenotypes in the F1 is different when comparing males and females, it would be tempting to suggest that the gene is X-linked. However, given that the reciprocal cross gives identical results suggests that the gene is autosomal. Seeing the different distribution between males and females one might consider sex-influenced inheritance as a model and have males more likely to express mahogany and females more likely to express red. This situation is similar to pattern baldness in humans. Consider two alleles which are autosomal and let

---

RR = red, Rr = red in females,

Rr = mahogany in males,

rr = mahogany.

P_1:

female: RR (red) X male: rr (mahogany)

F_1:

Rr = females red; males mahogany

    1/2 females (red)

    1/2 males (mahogany)

F_2:

    1/4 RR; 2/4 Rr, 1/4 rr

Because half of the offspring are males and half are females, one could, for clarity, rewrite the F_2 as:

	1/2 *females*	1/2 *males*
1/4 RR	1/8 red	1/8 red
2/4 Rr	2/8 red	2/8 mahogany
1/4 rr	1/8 mahogany	1/8 mahogany

33. In looking at the information provided in the text, notice that the only genotype which gives cock-feathering in males is hh, while three genotypes give hen-feathering in females,

    HH, Hh, and hh.

Remember that these genes are sex-limited and autosomal.

P_1:   female: HH X   male: hh

F_1:

    all hen-feathering

F_2:

	1/2 females	1/2 males
1/4 HH	hen-feathering	hen-feathering
2/4 Hh	hen-feathering	hen-feathering
1/4 hh	hen-feathering	cock-feathering

All of the offspring would be hen-feathered except for 1/8 males which are cock-feathered.

34. Passage of X-linked genes typically occurs from carrier mother to effected son. The fact that the father in couple #2 has hemophilia would not predispose his son to hemophilia. The #1 couple has no valid claim.

35. Phenotypic expression is dependent on the genome of the organism, the immediate molecular and cellular environment of the genome, and numerous interactions between a genome, the organism, and the environment.

36. *Penetrance* refers to the percentage of individuals which express the mutant phenotype while *expressivity* refers to the range of expression of a given phenotype.

37. *Anticipation* occurs when a heritable disorder exhibits a progressively earlier age of onset and an increased severity in successive generations. *Imprinting* occurs when phenotypic expression is influenced by the parental origin of the chromosome carrying a particular gene.

38. First, look for familiar ratios which will inform you as to the general mode of inheritance. Notice that the last cross (h) gives a 9:4:3 ratio which is typical of epistasis. From this information one can develop a model to account for the results given.

Symbolism:

$A\_B\_$ = black

A_bb = golden
$aabb$ = golden

$aaB\_$ = brown

The combination of bb is epistatic to the A locus.

(a) *AAB-* X *aaBB* (other configurations possible but each must give all offspring with A and B dominant alleles)

(b) *AaB-* X *aaBB* (other configurations are possible but no types can be produced)

(c) *AABb* X *aaBb*

(d) *AABB* X *aabb*

(e) *AaBb* X *Aabb*

(f) *AaBb* X *aabb*

(g) *aaBb* X *aaBb*

(h) *AaBb* X *AaBb*

Those genotypes which will breed true will be as follows:

black   = *AABB*

golden = all genotypes which are *bb*

brown = *aaBB*

39. It is probably best to simplify the problem by considering only the actual colors of canary, orange, red and salmon. There are three crosses to consider, each involving true-breeding strains. Because there are four different phenotypes possible, one might consider at least two interacting loci involved. All phenotypes involve color, so interaction between those two (or more) loci is probable occurring. **(a)** Notice that in Cross I, crossing canary with orange gives an all canary F_1 and a 9:3:4 ratio emerges in the F_2. This ratio suggests epistasis. Since canary represents the 9/16 class one might begin to think of the canary color being at the end of a pigment pathway.

$$\begin{array}{ll} aa & bb \end{array}$$

orange>>\|>red>\|>canary

9/16	$A\_B\_$	= canary
3/16	A_bb	= red
3/16	$aaB\_$	= orange
1/16	$aabb$	= orange

In looking at Cross II, notice that salmon behaves just like orange in crosses, so one might consider the following:

orange / salmon >>>> red>>>>canary

In order to account for the two colors, orange and salmon, let *A* exist in two mutant states a^s and a^s for generating the orange and salmon colors, respectively.

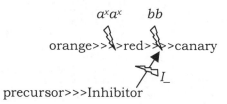

orange (a^o) ⎤
                ⎦ → red>>>>canary
salmon (a^s) ⎤

(b) When examining Cross III, another issue arises in that with a 9:7 ration one would have to "extend" the explanation as indicated in the problem. To do so, consider an inhibitory pathway driven by an "*I*" allelic pair.

$a^x a^x$ *bb*

orange>>↘>red>>↘>canary
                  ↖
                     I

precursor>>>Inhibitor

Notice that when the *I_* condition exists, the inhibitor's action is blocked and the red pigment is converted to canary. When the *ii* condition exists, the inhibitor functions and red pigment accumulates.

(c) While there are other possibilities, the following genotypes account for the results in a fairly efficient manner:

*A_B_I_*	=	canary
$a^o a^o$__	=	orange
$a^s a^s$__	=	salmon
A_Bii	=	red
*A_bb*__	=	red

(d) When testing the models presented above, one could conduct the following crosses:

Some pure-breeding red strains when crossed with some other pure-breeding red strains could yield canary as follows:

AABBii   X   *AAbbII*

⇓

AaBbIi (canary)

A second test of the model would be to cross any salmon with any orange strain. The results from such would never produce canary or red.

40. A first glance would seem to favor a 9:7 ratio, however, the phenotypes would have to be reversed for such a result to fit. Therefore, one must consider an alternative explanation. A 27:9:9:9:3:3:3:1 ratio fits very well with *A_B_C_* being purple and any homozygous recessive combination giving white. Thus a 27(purple):37(white) ratio fits well. To test this hypothesis, one might take the purple F_1's and cross them to the pure breeding (*aabbcc*) white type. Such a cross should give a 1(purple):7(white).

41. The clue to the solution comes from the description of the Dexters as not true-breeding and of low fertility. This indicates that Dexters are heterozygous and the Kerry breed is homozygous recessive. The homozygous dominant type is lethal. Polled is caused by an independently assorting dominant allele, while horned is caused by the recessive allele to polled.

42. (a) Because the denominator in the ratios is 64 one would begin to consider that there are three independently assorting gene pairs operating in this problem. Because there are only two characteristics (eye color and croaking) however, one might hypothesize that two gene pairs are involved in the inheritance of one trait while one gene pair is involved in the other.

(b) Notice that there is a 48:16 (or 3:1) ratio of rib-it to knee-deep and a 36:16:12 (or 9:4:3) ratio of blue to green to purple eye color. Because of these relationships one would conclude that croaking is due to one (dominant/recessive) gene pair while eye color is due to two gene pairs. Because there is a (9:4:3) ratio regarding eye color, some gene interaction (epistasis) is indicated.

(c,d) Symbolism:

Croaking: *R_* = rib-it; *rr* = knee-deep

Eye color:

Since the most frequent phenotype is blue eye, let *A_B_* represent the genotypes. For the purple class, "a 3/16 group" use the *A_bb* genotypes. The "4/16" class (green) would be the *aaB_* and the *aabb* groups.

(e) The cross involving a blue-eyed, knee-deep frog and a purple-eyed, rib-it frog would have the genotypes:

AABBrr   X   AAbbRR

which would produce an F₁ of *AABbRr* which would be blue-eyed and rib-it. The F2 will follow a pattern of a 9:3:3:1 ratio because of homozygosity for the A locus and heterozygosity for both the B and R loci.

9/16   *AAB_R_* = blue-eyed, rib-it

3/16   *AAB_rr* = blue-eyed, knee-deep

3/16   *AAbbR_*= purple-eyed, rib-it

1/16   *AAbbrr* = purple-eyed, knee-deep

(f) The different results can arise because of the genetic variety possible in producing the green-eyed frogs. Since there is no dependence on the B locus, the following genotypes can define the green phenotype:

aaBB, aaBb, aabb

(g) In doing these types of problems, take each characteristic individually, then build the complete genotypes. Notice that the ratio of purple-eyed to green-eyed frogs is 3:1, therefore expect the parents to be heterozygous for the *A* locus. Because the ratio of rib-it to knee-deep is also 3:1 expect both parents to be heterozygous at the R locus.

The *B* locus would have the *bb* genotype because both parents are purple-eyed as given in the problem. Both parents would therefore be *AabbRr*.

43. In looking at the pedigree, one can see that the typical carrier mother-to-son X-linked pattern is not present. One can't default to a Y-linked pattern because of sons (II-1, IV-1) not having precocious puberty. We might then consider a sex-limited form of inheritance where the gene(s) is(are) autosomal, but expression is limited to one sex, male in this case. Because of the relatively high frequency of occurrence of precocious puberty in the pedigree, one might consider a dominant gene being involved. Indeed, there is no skipping of generations typical of recessive traits. Notice however, that there is an apparent skipping of generations in giving rise to the IV-5 son. This is because females are not capable of expressing the gene. Given the degree of outcrossing, that the gene is probably quite rare and therefore heterozygotes are uncommon, and that the frequency of transmission is high, it is likely that this form of male precious puberty is caused by an autosomal dominant, sex-limited gene.

44. Given that both parents are true-breeding and the sort of gene interaction described is occurring, one can come up with the following symbols:

(a)   P₁: *YYBB*   X   *yWbb*

F₁: *YyBb* and *YWBb*

Crossing these F₁'s gives the observed ratios in the F₂.

(b) Given a blue male with the genotype *yyBb* and a green female with the genotype *YWBb*, the offspring are as given part **b** of the question.

45. (a) When the breeder crosses the F_1 brown animals, those which are *BB* X *BB* or *BB* X *Bb* will only produce brown offspring. Only those crosses which are *Bb* X *Bb* will produce white offspring. **(b)** Given that animals with white pelts (*bb*) breed poorly, the breeder would do well to only breed those brown animals which previously yielded white offspring: *Bb* X *Bb*.

46. Even though three gene pairs are involved, notice that because of the pattern of mutations, each cross may be treated as monohybrid **(a)** or dihybrid **(b,c)**.

(a) F_1:   *AABbCC =*   speckled

    F_2: 3   *AAB_CC =*   speckled
        1   *AAbbCC =*   yellow

(b) F_1:   *AABbCc =*   speckled

    F_2: 9   *AAB_C_ =*   speckled
        3   *AAB_cc =*   green
        3   *AAbbC_ =*   yellow $\Big\}$ 4
        1   *AAbbcc =*   yellow

(c) F_1:   *AaBBCc =*   speckled

    F_2: 9   *A_BBC_ =*   speckled
        3   *A_BBcc =*   green
        3   *aaBBC_ =*   colorless $\Big\}$4
        1   *aaBBcc =*   colorless

47. A cross of the following nature would satisfy the data:

    AABBCC   X   *aabbcc*

Offspring in the F_2:

27	*A_B_C_*	= purple
9	*A_B_cc*	= pink
9	*A_bbC_*	= rose
9	*aaB_C_*	= orange
3	*A_bbcc*	= pink
3	*aaB_cc*	= pink
3	*aabbC_*	= rose
1	*aabbcc*	= pink

           c   **b**   **a**
pink-✧> rose -✧> orange -✧> purple

The above hypothesis could be tested by conducting a backcross as given below:

    AaBbCc   X   *aabbcc*

The cross should give a

4(pink):2(rose):1(orange):1(purple) ratio

Chapter 11: Sex Determination and Sex Chromosomes

Concept Areas	Corresponding Problems
Sex Chromosomes	1, 3, 7, 8, 18
Sex Determination	2, 3, 4, 5, 14, 15, 17, 20, 21, 22
Sexual Differentiation	6, 16, 19, 20
Dosage Compensation	9, 10, 11, 12, 13, 23

Vocabulary: Organization and Listing of Terms and Concepts

Structures and Substances

Heteromorphic sex chromosomes

isogamete

zoospore

gametophyte

sporophyte

stamen (tassels)

microgametophyte

pistil

endosperm nuclei

oocyte nucleus

synergids

antipodal nuclei

heterochromosome

Y chromosome

heterogametic sex

homogametic sex

aromatase

Testis determining factor (TDF)

glucose-6-phosphate dehydrogenase deficiency (*G-6-PD*)

clone

X-inactivation center (*XIC*)

X-inactive specific transcript (*XIST*)

Xic, Xist (mouse)

open reading frame (ORF)

transformer gene (*tra*)

Sex-lethal (S*xl*)

double-sex (*dsx*)

maleless (*mle*)

Processes/ Methods

Sexual Differentiation

primary, secondary

unisexual

dioecious

gonochoric

bisexual

Chapter 11 *Sex Determination and Sex Chromosomes*

monoecious

hermaphroditic

intersex

Chlamydomonas

isogametes

Zea mays

double fertilization

C. elegans

XX/XO *Protenor* mode

XX/XY *Lygaeus* mode

ZZ/ZW

Sex determination (humans)

XX = female, XY = male

intersexuality

Klinefelter syndrome 47, XXY

48, XXXY

48, XXYY, *etc.*

Turner syndrome 45, X

mosaics 45X/46XY, 45X/46XX

47, XXX; 48, XXXX

49, XXXXX

47, XYY

sexual differentiation

gonadal primordia

cortex, medulla

human Y chromosome

pseudoautosomal regions (PARS)

NRY

testis determining factor (TDF)

sex determining region (SRY)

XX males

XY females

transgenic mice

SOX9, WT1, SF1

Sex ratio (humans)

primary

secondary

Dosage compensation

sex chromatin body (Barr body)

N-1 rule

Lyon hypothesis, Lyonization

G-6-PD

red-green color blindness

anhidrotic ectodermal dysplasia

X-inactivating center (*XIC*)

X-inactive specific transcript (*XIST*)

open reading frame (ORF)

epigenetic event

Chapter 11 Sex Determination and Sex Chromosomes

Sex determination (*Drosophila*)

non-disjunction

XO = sterile male

XXY = normal female

ratio (number of X chromosomes to number of haploid sets of autosomes)

superfemale (metafemale)

metamale

intersex

RNA splicing, alternative splicing

dosage compensation

mosaics, bilateral gynandromorph

Concepts

Sex determination (humans, *Drosophila*)

Sex differentiation

Dosage compensation

Solutions to Problems and Discussion Questions

1. The term *heteromorphic* refers to the condition in many organisms where there are two different forms (morphs) of chromosomes such as X and Y. *Heterogamy* refers to the condition where there are two different sizes of gametes such as egg and sperm.

2. The *Protenor* form of sex determination involves the XX/XO condition while the *Lygaeus* mode involves the XX/XY condition.

3. Calvin Bridges (1916) studied nondisjunctional *Drosophila* which had a variety of sex chromosome complements. He noted that XO produced sterile males while XXY produced fertile females. The Y chromosome is male determining in humans. Individuals with the 47,XXY complement are males while 45,XO produces females.

In *Drosophila* it is the balance between the number of X chromosomes and the number of haploid sets of autosomes which determines sex. In humans there is a small region on the Y chromosome which determines maleness.

4. In *primary* nondisjunction half of the gametes contain two X chromosomes while the complementary gametes contain no X chromosomes. Fertilization, by a Y-bearing sperm cell, of those female gametes with two X chromosomes would produce the XXY Klinefelter syndrome. Fertilization of the "no-X" female gamete with a normal X-bearing sperm will produce the Turner syndrome.

5. (a) female $X^{rw}Y$ X male X^+X^+

F_1: females: X^+Y (normal)
      males: $X^{rw}X^+$ (normal)

F_2: females: X^+Y (normal)
               $X^{rw}Y$ (reduced wing)
      males: $X^{rw}X^+$ (normal)
               X^+X^+ (normal)

(b) female $X^{rw}X^{rw}$ X   male X^+Y

F_1: females: $X^{rw}X^+$ (normal)
      males: $X^{rw}Y$ (reduced wing))

F_2: females: $X^{rw}X^+$ (normal)
               $X^{rw}X^{rw}$ (reduced wing)
      males: X^+Y (normal)
               $X^{rw}Y$ (reduced wing)

(c) No.

6. Males and females share a common placenta and therefore hormonal factors carried in blood. Hormones and other molecular species (transcription factors perhaps) triggered by the presence of a Y chromosome lead to a cascade of developmental events which both suppress female organ development and enhance mastulinization. Other mammals also exhibit a variety of similar effects depending on the sex of their uterine neighbors during development.

7. Because attached-X chromosomes have a mother-to-daughter inheritance and the father's X is transferred to the son, one would see daughters with the white eye phenotype and sons with the miniature wing phenotype.

8. Because synapsis of chromosomes in meiotic tissue is often accompanied by crossing over, it would be detrimental to sex-determining mechanisms to have sex-determining loci on the Y chromosome transferred, through crossing over, to the X chromosome.

9. A *Barr body* is a darkly staining chromosome seen in some interphase nuclei of mammals with two X chromosomes. There will be one less Barr body than number of X chromosomes. The Barr body is an X chromosome which is considered to be genetically inactive.

10. There is a simple formula for determining the number of Barr bodies in a given cell: N-1, where N is the number of X chromosomes.

Klinefelter syndrome (XXY)	= 1
Turner syndrome (XO)	= 0
47, XYY	= 0
47, XXX	= 2
48, XXXX	= 3

11. The *Lyon Hypothesis* states that the inactivation of the X chromosome occurs at random early in embryonic development. Such X chromosomes are in some way "marked" such that all clonally-related cells have the same X chromosome inactivated.

12. Females will display mosaic retinas with patches of defective color perception. Under these conditions, their color vision may be influenced.

13. Refer to the K/C text and notice that the phenotypic mosaicism is dependent on the heterozygous condition of genes on the two X chromosomes. Dosage compensation and the formation of Barr bodies occurs only when there are two or more X chromosomes. Males normally have only one X chromosome therefore such mosaicism can not occur. Females normally have two X chromosomes. There are cases of male calico cats which are XXY.

14. Many organisms have evolved over millions of years under the fine balance of numerous gene products. Many genes required for normal cellular and organismic function in *both* males and females are located on the X chromosome. These gene products have nothing to do with sex determination or sex differentiation.

15. In mammals, the scheme of sex determination is dependent on the presence of a piece of the Y chromosome. If present a male is produced. In *Bonellia viridis*, the female proboscis produces some substance which triggers a morphological, physiological, and behavioral developmental pattern which produces males.

To elucidate the mechanism, one could attempt to isolate and characterize the active substance by testing different chemical fractions of the proboscis. Secondly, mutant analysis usually provides critical approaches into developmental processes. Depending on characteristics of the organism, one could attempt to isolate mutants which lead to changes in male or female development. Third, by using micro-tissue transplantations, one could attempt to determine which "centers" of the embryo respond to the chemical cues of the female.

16. There are several possibilities which are discussed in the K/C text. One could account for the significant departures from a 1:1 ratio of males to females by suggesting that at anaphase I of meiosis, the Y chromosome more often goes to the pole which produces the more viable sperm cells. One could also speculate that the Y-bearing sperm has a higher likelihood of surviving in the female reproductive tract, or that the egg surface is more receptive to Y-bearing sperm. At this time the mechanism is unclear.

17. Since there is a region of synapsis close to the SRY-containing section on the Y chromosome, crossing over in this region would generate XY translocations which would lead to the condition described.

18. Because of the homology between the *red* and *green* genes, there exists the possibility for an irregular synapsis (see the figure below) which, following crossing over, would give a chromosome with only one (*green*) of the duplicated genes. When this X chromosome combines with the normal Y chromosome, the son's phenotype can be explained.

"Normal Synapsis"

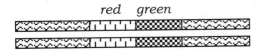

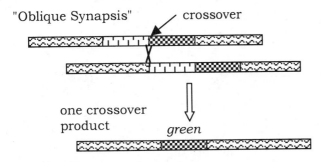

"Oblique Synapsis"   crossover

one crossover
product   *green*

19. The presence of the Y chromosome provides a factor (or factors) which leads to the initial specification of maleness. Subsequent expression of secondary sex characteristics must be dependent on the interaction of the normal X-linked *Tfm* allele with testosterone. Without such interaction, differentiation takes the female path.

20. (a) Something is missing from the male-determining system of sex determination either at the level of the genes, gene products, or receptors, *etc.*.

(b) The *SOX9* gene or its product is probably involved in male development. Perhaps it is activated by *SRY*.

(c) There is probably some evolutionary relationship between the *SOX9* gene and *SRY*. There is considerable evidence that many other genes and pseudogenes are also homologous to *SRY*.

(d) Normal female sexual development does not require the *SOX9* gene or gene product(s).

(e) *SRY* may activate *SOX9*, which is also required for normal skeletal development. A good review of SRY and its relatives among a variety of vertebrates is found in Jeyasuria and Pace. 1998. *Journal of Experimental Zoology* 281:428-449

21. Since all haploids are male and half of the egges are unfertilized, 50% of the offspring would be male at the start; adding the X_a/X_a types gives 25% more male, the remainder X_a/X_b would be female. Overall, 75% of the offspring would be male while 25% would be female.

22. In snapping turtles, sex determination is strongly influenced by temperature such that males are favored in the 26-34 °C range. Lizards, on the other hand, appear to have their sex determined by factors other than temperature in the 20-40 °C range.

23. We can assume that the donor cat "Rainbow" was heterozygous for the *black* and *orange* genes and that both genes were present in CC since a diploid nucleus was used. The white patches of CC are due to an autosomal gene *S* for white spotting which prevents pigment formation in the cell lineages in which it is expressed. Homoygous *SS* cats have more white than heterozygous *Ss* cats. The absence of orange patches is due to the fact that in gonadal tissue, while oogonia have a single active X chromosome, the inactive X chromosome is reactivated at, or more likely, shortly before, entry into meiotic prophase (Kratzer and Chapman 1981). Thus X chromosome inactivation does not remain in certain ovarian cells as in somatic tissue. With *black* being expressed in the presence of orange, only black shows through in the Carbon Copy's coat. CC can be seen at

http://cats.about.com/library/weekly/aa021502a.htm.

Chapter 12: Linkage, Crossing Over, and Mapping in Eukaryotes

Concept Areas	Corresponding Problems
Linkage vs. Independent Assortment	1, 26, 38
Chromosome Mapping	3, 4, 5, 6, 7, 8, 9, 10, 11, 12, 13, 25, 26, 27 28, 29, 31, 32, 33, 34, 36, 37, 20, 24, 30, 35 , 38
Multiple Crossovers and Three-Point Mapping	5, 14, 15, 16, 17, 18, 19
Determining Gene Sequence	15, 16, 17, 20
Interference and Coefficient of Coincidence	6, 15, 16
Crossing Over in the Four-Strand Stage	2
Mechanism of Crossing Over	2, 18, 23
Mitotic Recombination	21, 22
Somatic Cell Hybridization and Human Maps	34

Vocabulary: Organization and Listing of Terms and Concepts

Structures and Substances

Chromosome map

    interlocus distance

    linkage group

    isogamete

    tetrad

    parental ditype

    nonparental ditype

    tetratype

Heterokaryon

Synkaryon

Ascospores

Ascus (pl. asci)

Rh antigen

Elliptocytosis

Huntington disease

Cystic fibrosis

Neurofibromatosis

Bromodeoxyuridine (BUdR)

DNA helicase

Processes/Methods

Linkage

    crossing over

    crossover gametes (recombinant)

      recombination

      reciprocal classes

    parental (noncrossover gametes)

Chapter 12 Linkage, Crossing Over, and Mapping in Eukaryotes

incomplete

chiasmata (chiasma)

three-point mapping

    product rule (multiple crossovers)

        noncrossovers (NCO)

        single crossovers (SCO)

        double crossovers (DCO)

Linkage ratio

Determining gene sequence

    correct heterozygous arrangement

    correct sequence of genes

        method I

        method II

    Poisson distribution

        2-strand double

        3-strand double

        4-strand double

           mapping function

Cytological evidence (crossing over)

Mechanism of crossing over

    synaptonemal complex

    zygotene DNA synthesis

    pachytene DNA synthesis

Human chromosome maps

    lod score

Somatic cell hybridization

    random loss of human chromosomes

    synteny testing

    translocation

Haploid organisms

    tetrad analysis

Mapping to the centromere

    first-division segregation

    second-division segregation

Gene conversion

Mitotic recombination

    synapsis

    genetic exchange

    twin spots

    parasexual cycle

Sister chromatid exchange

    bromodeoxyuridine (BUdR)

    Bloom syndrome

Mendel and linkage

    independent assortment

Concepts

Linked genes (linkage groups)

    arrangement (gene sequence)(F12.1)

Chromosome Theory of Inheritance

Chromosome maps	expected frequency of DCO
map unit (% recombination)	observed frequency of DCO
50% maximum	positive
Chiasmatype theory	negative
Interference	Generation of variation
coefficient of coincidence	Mitotic recombination

F12.1 Illustration of critical arrangements of linked genes. Notice that there are two possible arrangements for a *AaBb* double heterozygote. In order to do linkage problems correctly, such arrangements must be understood.

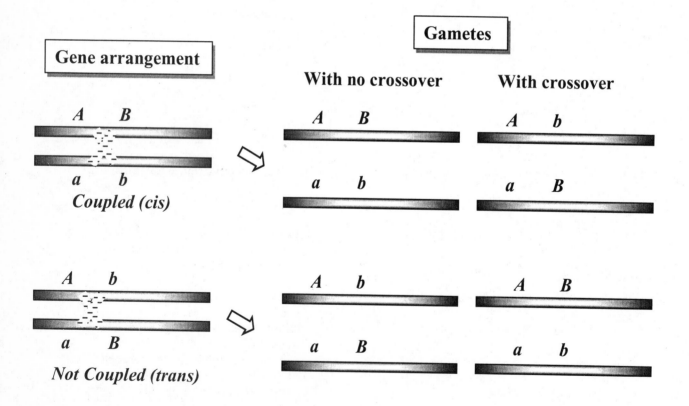

Solutions to Problems and Discussion Questions

1. The biological significance of genetic exchange and recombination appears to be to generate genetic variation in gametes, thereby leading to genetic variation in organisms. By reshuffling genes, new combinations are generated which may then be of evolutionary advantage. In addition, because chromosomal position can influence gene function, variation is created by *position effect*.

2. First, in order for chromosomes to engage in crossing over, they must be in proximity. It is likely that the side-by-side pairing which occurs during synapsis is the earliest time during the cell cycle that chromosomes achieve that necessary proximity. Second, chiasmata are visible during prophase I of meiosis and it is likely that these structures are intimately associated with the genetic event of crossing over.

3. With some qualification, especially around the centromeres and telomeres, one can say that crossing over is somewhat randomly distributed over the length of the chromosome. Two loci which are far apart are more likely to have a crossover between them than two loci that are close together.

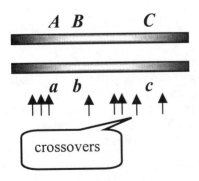

4. Because crossing over occurs at the four-strand stage of the cell cycle (that is, after S phase) notice that each single crossover involves only two of the four chromatids.

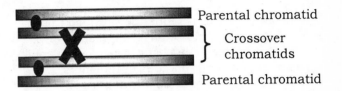

5. As mentioned in an earlier answer (#3) with some qualifications, crossovers occur randomly along the lengths of chromosomes. Within any region, the occurrence of two events is less likely than the occurrence of one event. If the probability of one event is

$$1/X,$$

the probability of two events occurring at the same time will be

$$1/X^2.$$

6. Positive interference occurs when a crossover in one region of a chromosome interferes with crossovers in nearby regions. Such interference ranges from zero (no interference) to 1.0 (complete interference). Interference is often explained by a physical rigidity of chromatids such that they are unlikely to make sufficiently sharp bends to allow crossovers to be close together.

7. Each cross must be setup in such a way as to reveal crossovers because it is on the basis of crossover frequency that genetic maps are developed. It is necessary that genetic heterogeneity exist so that different arrangements of genes, generated by crossing over, can be distinguished.

The organism which is heterozygous must be the sex in which crossing over occurs. In other words, it would be useless to map genes in *Drosophila* if the male parent is the heterozygote since crossing over is not typical in *Drosophila* males.

Lastly, the cross must be setup so that the phenotypes of the offspring readily reveal their genotypes. The best arrangement is one where a fully heterozygous organism is crossed with an organism which is fully recessive for the genes being mapped.

8. Since the distance between *dp* and *ap* is greatest, they must be on the "outside" and *cl* must be in the middle. The genetic map would be as follows:

$$dp\text{---}cl\text{----------------------}ap$$

    3 *mu.*   39 *mu.*

9. The initial cross for this problem would be

    AaBb   X   *aabb*.

(a) If the two loci are on different chromosomes, independent assortment would occur and the following distribution (1:1:1:1) is expected:

1/4	*AaBb*
1/4	*Aabb*
1/4	*aaBb*
1/4	*aabb*

(b) Even though the two loci are linked and on the same chromosome, the frequency of crossing over is so high that crossovers always occur. Under that condition independent assortment would occur and the following distribution (1:1:1:1) is expected:

1/4	*AaBb*
1/4	*Aabb*
1/4	*aaBb*
1/4	*aabb*

(c) If crossovers never occur, then all of the gametes from the heterozygous parent are *parental*. If the arrangement is

    AB/ab   X   *ab/ab*

then the two types of offspring will be

    1/2   *AB/ab*

    1/2   *ab/ab*.

Under this condition *AB* are *coupled*. If, however, *A* and *B* are not coupled then the symbolism would be

    Ab/aB   X   *aabb*.

The offspring would occur as follows:

    1/2   *Ab/ab*

    1/2   *aB/ab*.

(d) If the loci are linked with 10 map units between them, then the two recombinant classes must add up to 10% of the total. Assuming that *A* and *B* are coupled, the following distribution would occur:

45%	*AaBb*	(parental)
5%	*Aabb*	(crossover)
5%	*aaBb*	(crossover)
45%	*aabb*	(parental)

10. In looking at this problem one can immediately conclude that the two loci (kernel color and plant color) are linked because the test cross progeny occur in a ratio other than 1:1:1:1 (and epistasis does not appear because all phenotypes expected are present).

The question is whether the arrangement in the parents is *coupled*

$$RY/ry \qquad X \qquad ry/ry$$

or *not coupled*

$$Ry/rY \qquad X \qquad ry/ry$$

Notice that the most frequent phenotypes in the offspring, the parentals, are colored, green (88) and colorless, yellow (92). This indicates that the heterozygous parent in the test cross is coupled

$$RY/ry \qquad X \qquad ry/ry$$

with the two dominant genes on one chromosome and the two recessives on the homologue (F6.1). Seeing that there are 20 crossover progeny among the 200, or 20/200, the map distance would be 10 map units (20/200 X 100 to convert to percentages) between the *R* and *Y* loci.

11. Start this problem by working through the expected offspring under two models. One with no crossing over and the second with 30% crossing over in the female.

No crossing over:

Female gametes: *Male gametes:*

$1/2 \ e \ ca^+$ $1/2 \ e \ ca^+$

$1/2 \ e^+ \ ca$ $1/2 \ e^+ \ ca$

Offspring:

1/4   "e" phenotype

2/4   wild

1/4   "ca" phenotype

With 30% crossing over.

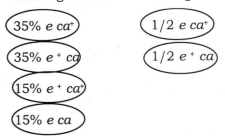

Female gametes: Male gametes:

$35\% \ e \ ca^+$ $1/2 \ e \ ca^+$

$35\% \ e^+ \ ca$ $1/2 \ e^+ \ ca$

$15\% \ e^+ \ ca^+$

$15\% \ e \ ca$

Offspring: (obtained by combining gametes and phenotypes)

"e" phenotype   = 17.5% + 7.5%   **= 25%**

wild phenotype = 17.5% + 7.5% + 17.5% + 7.5%
                                = 50%

"ca" phenotype = 17.5% + 7.5%   **= 25%**

Notice that the distribution of phenotypes is the same, regardless of the contribution of the crossover classes.

12. Since there is no indication as to the configuration of the *P* and *Z* genes (*coupled or not coupled*) in the parent, one must look at the percentages in the offspring. Notice that the most frequent classes are *PZ* and *pz*. These classes represent the parental (non-crossover) groups which indicates that the original parental arrangement in the test cross was

$$PZ/pz \qquad X \qquad pz/pz$$

Adding the crossover percentages together (6.9 + 7.1) gives 14% which would be the map distance between the two genes.

13. This problem can be approached by looking for the most distant loci (*adp* and *b*) then filling in the intermediate loci. In this case the map for parts **(a)** and **(b)** is the following:

> *d*...........*b*.........*pr*..........*vg*........*c*.........*adp*
> 31   48   54   67   75   83
>
> **Map Units**

The expected map units between *d* and *c* would be 44, *d* and *vg* would be 36, and *d* and *adp* 52. However because there is a theoretical maximum of 50 map units possible between two loci in any one cross, that distance would be below the 52 determined by simple subtraction.

14.

	female A:	*female* B:	*Frequency:*
NCO	3, 4	7, 8	first
SCO	1, 2	3, 4	second
SCO	7, 8	5, 6	third
DCO	5, 6	1, 2	fourth

The single crossover classes which represent crossovers between the genes which are closer together (*d-b*) would occur less frequently than the classes of crossovers between more distant genes (*b-c*).

15. For two reasons, it is clear that the genes are in the *coupled* configuration in the F$_1$ female. First a completely homozygous female was mated to a wild type male and second, the phenotypes of the offspring indicate the following parental classes

> *sc s v* and + + +

(a)

P$_1$:

> *sc s v / sc s v* X + + +/Y

F$_1$:

> + + +/*sc s v* X *sc s v*/Y

(b) Using Methods I or II for determining the sequence of genes, examine the parental classes and compare the arrangement with the double crossover (least frequent) classes. Notice that the *v* gene "switches places" between the two groups (parentals and double crossovers). The gene which switches places is in the middle.

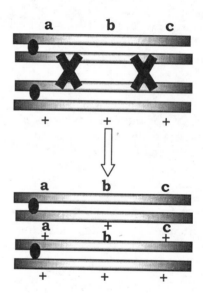

The map distances are determined by first writing the proper arrangement and sequence of genes, then computing the distances between each set of genes.

> *sc v s*
> + + +

sc - v = $\dfrac{150 + 156 + 10 + 14}{1000}$ X 100

= 33% (map units)

v - s = $\dfrac{46 + 30 + 10 + 14}{1000}$ X 100

= 10% (map units)

Double crossovers are always added into each crossover group because they represent a crossover in each region.

$s\underbrace{c\text{-------}v}_{33}\underbrace{\text{-----}s}_{10}$

(c) The coefficient of coincidence =

$\dfrac{\text{observed freq. double C/O}}{\text{expected freq. double C/O}}$

= $\dfrac{(14 + 10)/1000}{.33 \text{ X } .1}$

= $\dfrac{.024}{.033}$

= .727

which indicates that there were fewer double crossovers than expected, therefore positive chromosomal interference is present.

16. This set-up involves an F_1 in which the fully heterozygous female has the genes y and w in *coupled* and *ct not coupled*. The arrangement for the cross is therefore:

(a) $y \, w + / + + ct$ X $y \, w + / Y$

It is important at this point to determine the gene sequence. Using Methods I or II, examine the parental classes and compare the arrangement with the double crossover (least frequent) classes. Notice that the w gene "switches places" between the two groups (parentals and double crossovers). The gene which switches places is in the middle.

Therefore the arrangement as written above is correct.

(b)

y - w = $\dfrac{9 + 6 + 0 + 0}{1000}$ X 100

= 1.5 map units

w - ct = $\dfrac{90 + 95 + 0 + 0}{1000}$ X 100

= 18.5 map units

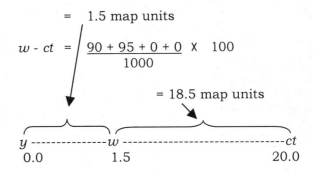

y --------------- w ----------------------------- ct
0.0   1.5   20.0

(c) There were

.185 X .015 X 1000 = 2.775

double crossovers expected.

(d) Because the cross to the F_1 males included the normal (wild type) gene for *cut wings* it would not be possible to unequivocally determine the genotypes from the F_2 phenotypes for all classes.

17. (a) The cross will be as follows. Represent the *Dichete* gene as an upper-case letter because it is dominant.

P_1:	$D + +/ + + +$	X	$+ e \, p/+ e \, p$
F_1:	$D + +/+ e \, p$	X	$+ e \, p/+ \, e \, p$
F_2:	$D + +/+ e \, p$	Dichete	
	$+ e \, p /+ e \, p$	ebony, pink	
	$D \, e +/+ e \, p$	Dichete, ebony	
	$+ + p/+ e \, p$	pink	
	$D + p/+ e \, p$	Dichete, pink	
	$+ e +/+ e \, p$	ebony	
	$D \, e \, p/+ e \, p$	Dichete, ebony, pink	
	$+ + +/+ e \, p$	wild type	

(b) Determine which gene is in the middle by comparing the parental classes with the double crossover classes. Notice that the *pink* gene "switches places" between the two groups (parentals and double crossovers). The gene which switches places is in the middle. So rewriting the sequence of genes with the correct arrangement gives the following:

F₁:

$D + +/+ p\ e$ X $+ p\ e/+ p\ e$

Distances: remember to add in the double crossover classes

$D\text{-}p = \dfrac{12 + 13 + 2 + 3}{1000} \times 100$

= 3.0 map units

$p\text{-}e = \dfrac{84 + 96 + 2 + 3}{1000} \times 100$

= 18.5 map units

18. The fact that two of the genes are linked and 20 map units apart on the third chromosome, and one is on the second chromosome, the problem is a combination of linkage and independent assortment. First provide the genotypes of the parents in the original cross and the reciprocal. Use a semicolon to indicate that two different chromosome pairs are involved.

P₁:

females: $+/+;\ p\ e/p\ e$

X

males: $dp/dp;\ + +/+ +$

F₁:

females: $+/dp;\ + +/p\ e$

X

males: $dp/dp;\ p\ e/p\ e$

Female gametes: use a modification of the forked-line method for determining the types of gametes to be produced. The *dumpy* locus will give .5 + and .5 *dp* to the gametes because of independent assortment (on a different chromosome) and the other two loci will segregate with 20% (map units) being the recombinants, and 80% being the parentals.

0.5 +
- 0.4 + + (parental) = 0.20 + + +
- 0.1 + e (crossover) = 0.05 + + e
- 0.1 p + (crossover) = 0.05 + p +
- 0.4 p e (parental) = 0.20 + p e

0.5 dp
- 0.4 + + (parental) = 0.20 dp + +
- 0.1 + e (crossover) = 0.05 dp + e
- 0.1 p + (crossover) = 0.05 dp p +
- 0.4 p e (parental) = 0.20 dp p e

Crossed with *dp p e* from the male gives the following offspring:

0.20 wild type
0.05 ebony
0.05 pink
0.20 pink, ebony
0.20 dumpy
0.05 dumpy, ebony
0.05 dumpy, pink
0.20 dumpy, pink, ebony

For the reciprocal cross:

F₁:

males: $+/dp$; $++/p\ e$

X

females: dp/dp; $p\ e/p\ e$

there would be no crossover classes.

0.5 + ⟋ 0.5 +   + (parental) = 0.25 + + +

0.5 + —— 0.5 p   e (parental) = 0.25 + p e

0.5 + ⟋ 0.5 +   + (parental) = 0.25 dp + +

0.5 dp —— 0.5 p   e (parental) = 0.25 dp p e

Crossed with *dp p e* from the female gives the following offspring:

.25 wild type

.25 pink, ebony

.25 dumpy

.25 dumpy, pink, ebony

The results would change because of no crossing over in males.

19. Since *Stubble* is a dominant mutation (and homozygous lethal) one can determine whether it is heterozygous ($Sb/+$) or homozygous wild type ($+/+$). One would use the typical test cross arrangement with the *curled* gene so the arrangement would be

$+ cu/ + cu$

20. In typical trihybrid crosses one expects eight kinds of offspring. In this example, only six are listed and one can assume that since the double crossover class is the least frequent, it is the double crossovers which are not listed.

To work this type of problem, examine the list to see which types are not present. In this case, the double crossover classes are the following:

$+ + c$ and $a\ b\ +$

(a,b) Notice that if you compare the parental classes (most frequent) with the double crossover classes (zero in this case) one can, by using the logic of the methods described in the text, determine that the gene *b* is in the middle and the arrangement is as follows. Note: for consistency the zeros (double crossovers) are included in the calculations.

$+ b\ c/ a + +$

$a - b\ = \dfrac{32 + 38 + 0 + 0}{1000} \times 100$

$= 7$ map units

$b - c\ = \dfrac{11 + 9 + 0 + 0}{1000} \times 100$

$= 2$ map units

(c) The progeny phenotypes that are missing are $+ + c$ and $a\ b +$, which, of 1000 offspring, 1.4 (.07 X .02 X 1000) would be expected. Perhaps by chance or some other unknown selective factor, they were not observed.

21. You will notice that it would take two crossovers to "isolate" the *singed* gene for a spot, and one exchange for a twin spot. Therefore the likelihood of a twin spot is greater than the likelihood of a singed spot. The arrangement of the "markers" being discussed in the second part of this question is as follows:

```
-----------t--------------------f---------c-------
          27.5                 56.7      66
```

Because the distance between *forked* and *tan* loci is relatively great, crossovers in this region would be most frequent (and produce a tan spot). The region between the centromere and *forked* is approximately 9 map units, therefore one would expect crossovers to occur in this region at the next highest frequency thereby yielding twin spots. The least frequent event would probably be forked spots because it would take two crossovers to "isolate" the *forked* gene.

22. Because sister chromatids are genetically identical (with the exception of rare new mutations) crossing over between sisters provides no increase in genetic variability. Individual genetic variability could be generated by somatic crossing over because certain patches on the individual would be genetically different from other regions. This variability would be of only minor consequence in all likelihood. Somatic crossing over would have no influence on the offspring produced.

23. These observations as well as the results of other experiments indicate that the synaptonemal complex is required for crossing over.

24. (a) There would be $2^n = 8$ genotypic and phenotypic classes and they would occur in a 1:1:1:1:1:1:1:1 ratio.

(b) There would be two classes and they would occur in a 1:1 ratio.

(c) There are 20 map units between the *A* and *B* loci and locus *C* assorts independently from both *A* and *B* loci.

25. Since the genetic map is more accurate when relatively small distances are covered and when large numbers of offspring are scored, this map would probably not be too accurate with such a small sample size.

26. Assign the following symbols for example:

R = Red	*r* = yellow
O = Oval	*o* = long

Progeny A: *Ro/rO* X *rroo* = 10 map units
Progeny B: *RO/ro* X *rroo* = 10 map units

27. The easiest way to approach this problem is to set up fractions representing the proportions of gametes, with the frequency of the recombinant gametes adding up to 25%. For each, the gamete proportions would be the following:

3/8 *Ab*; 3/8 *aB*; 1/8 *AB*; 1/8 *ab*

Now, combine the gametes from each parent (they are the same) and arrive at the following frequency:

*A_B_* 33/64; *A_bb* 15/64; *aaB_* 15/64; *aabb* 1/64

28. The map distance of a gene to the centromere in *Neurospora* is determined by dividing the percentage of second division asci (tetrads) by two. Patterns other than *BBbb* or *bbBB* are "second division" as discussed in the text and represent a crossover between the gene in question and the centromere. In the data given, the percentage of second division segregation is 20/100 or 20%. Dividing by 2 (because only two of the four chromatids are involved in any single crossover event) gives 10 map units.

29. Because the two types of tetrads occurred at equal frequency, one could say that the gene loci are not linked. Also, looking at the individual loci, notice that gene a segregates just as often with gene *b* as it does with its allele +. Because there are no arrangements characteristic of second division segregation (*a+a+*, or *b++b*, for example) the two genes must be very close to their centromeres.

30. The general formula for determination of map distances is as follows:

$$Map\ distance = \frac{NP + 1/2(T)}{Total\ \#\ Tetrads}$$

NP = Nonparental ditypes
T = Tetratypes

For Cross 1:

$$\frac{36 + 14}{100} = 50\ map\ units$$

Because there are 50 map units between genes *a* and *b*, they are not linked.

For Cross 2:

$$\frac{3 + 9}{100} = 12\ map\ units$$

Because genes *a* and *b* are not linked they could be on non-homologous chromosomes or far apart (50 map units or more) on the same chromosome. Because genes *c* and *b* are linked and therefore on the same chromosome, it is also possible that genes *a* and *c* are on different chromosome pairs. Under that condition, the NP and P (parental ditypes) would be equal; however, there is a possibility that the following arrangement occurs and that genes *a* and *c* are linked.

```
        <50           12
a---------------c-----------b
        _____/
              >50
```

31.

(a)

Tetrad	Category
1	parental ditype
2	parental ditype
3	nonparental ditype
4	tetratype
5	tetratype
6	tetratype

(b) If a single crossover occurs between the centromere and the two linked genes, then the arrangement in tetrad 2 will occur.

(c)

$$Map\ distance = \frac{NP + 1/2(T)}{Total\ \#\ Tetrads}$$

NP = Nonparental ditypes
T = Tetratypes

$$= \frac{6 + 11}{71} = about\ 24\ map\ units$$

32. (a) One can see the various arrangements and crossover events which are labeled as parental ditype (P), nonparental ditype (NP), and tetratype (T). From that information, the following can be listed:

Tetrad in Problem	Class
1	NP
2	T
3	P
4	NP
5	T
6	P
7	T

(b) The easiest way to determine whether the *c* and *d* genes are linked is to compare the frequencies of the parental ditype (P) and nonparental ditype (NP) classes. If P>NP, then the genes are linked. If P=NP they are independently assorting. In the problem given here, P = 44 and NP = 2. Therefore the genes are linked.

(c) The gene to centromere distances are computed by dividing the percent second division segregation by two.

For the centromere to *c* distance:

$$\frac{1 + 5 + 3 + 1}{69} \times 100 = 14.5$$

now divide by 2 = 7.2 map units

For the centromere to *d* distance:

$$\frac{17 + 1 + 3 + 1}{69} \times 100 = 31.9$$

now divide by 2   = 15.9 map units

The map would be, according to these figures:

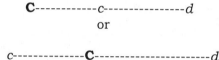

By mapping the genes to the centromeres, one can come up with two configurations. Either the genes are on the same side of the centromere or they are on opposite sides. To decide which configuration is occurring, one should see what type of crossovers are required in each case. The two tetrad arrangements which are critical in dealing with the above configurations are #4 and #5.

Notice that, for tetrad arrangement #4 it takes three crossovers if genes *c* and *d* are on the same side of the centromere while it takes two crossovers if they are on opposite sides. For tetrad arrangement #5 it takes two crossovers if genes *c* and *d* are on the same side of the centromere while it takes three crossovers if they are on opposite sides. Notice that the frequency of tetrad arrangement #5 (5) is much greater than tetrad arrangement #4 (1). This would make sense if the genes were on the same side of the centromere because the highest number of tetrads (5) is in tetrad arrangement #5 where the lowest number of crossovers is required.

If the two genes were on opposite sides of the centromere, the highest number of tetrad arrangements (5 in #5) would be associated with three crossover events. Since the likelihood of multiple crossovers decreases as the number of crossovers increases, it would seem reasonable that the genes are on the same side of the centromere.

Tetrad class #4

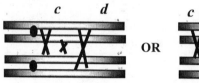 OR

Tetrad class #5

 OR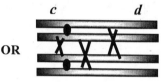

(d) The formula for calculating the distance between the two loci is as follows:

$$\text{Map distance} = \frac{NP + 1/2(T)}{\text{Total \# Tetrads}}$$

NP =   Nonparental ditypes
T   =   Tetratypes

$$= \frac{2 + 12}{69} = 20 \text{ map units}$$

(e)

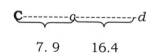

7. 9   16.4

The discrepancy between the two mapping systems is caused by the manner in which first and second division segregation products are scored. For instance, in tetrad arrangement #1 there are actually two crossovers between the *d* gene and the centromere, but it is still scored as a first division segregation. In tetrad arrangement #4, three crossovers occur between the *d* gene and the centromere but they are scored as one. If one draws out all the crossovers needed to produce the tetrad arrangements in this problem it would become clear that there are many crossovers between the *d* gene and the centromere which go undetected in the scoring of the arrangements of the *d* gene itself. This will cause one to underestimate the distance and give the discrepancy noted. One could account for these additional crossover classes to make the map more accurate.

(f)

Tetrad class #6

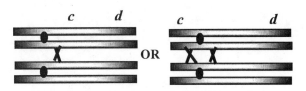

OR

33. (a)

10 *AB, Ab, aB, ab* (tetratype)
102 *Ab, aB, Ab, aB* (nonparental ditype)
99 *AB, AB, ab, ab* (parental ditype)

(b) Since the Parental Ditype class is approximately equal to the Nonparental Ditype class, one would conclude that there is no linkage.

(c) Since there is no linkage, one would conclude that the two loci are either on nonhomologous chromosomes or far apart on homologous chromosomes.

34. Look for overlap between chromosome number in given clones and genes expressed. Note that *ENO1* is expressed in clones B,D,E; chromosomes 1 and 5 are common to these clones. However, since *ENO1* is not expressed in clone C which is missing chromosome 1 (and has chromosome 5), *ENO1* must be on chromosome 1.

MDH1: chromosome 2
PEPS: chromosome 4
PMG1: chromosome 1

35. First make a drawing with the genes placed on the homologous chromosomes as follows:

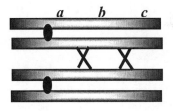

Realize that there are four chromatids in each tetrad and a single crossover involves only two of the four chromatids. Non-involved chromatids must be added to the non-crossover classes. Do all the crossover classes first, then add up the non-crossover chromatids. For example, in the first crossover class (20 between *a* and *b*) notice that there will be 40 chromatids which were not involved in the crossover. These 40 must be added to the *abc* and +++ classes.

a b c	=	168
+ + +	=	168
a + +	=	20
+ *b c*	=	20
+ + *c*	=	10
a b +	=	10
+ *b* +	=	2
a + *c*	=	2

Chapter 12 Linkage, Crossing Over, and Mapping in Eukaryotes

The map distances would be computed as follows:

$$a - b = \frac{20 + 20 + 2 + 2}{400} \times 100$$

$$= 11 \text{ map units}$$

$$b - c = \frac{10 + 10 + 2 + 2}{400} \times 100$$

$$= 6 \text{ map units}$$

36. It is important to remember that there is no crossing over in males and that if two genes are on the same chromosome, there will be complete linkage of the genes in the male gametes. In females, crossing over will produce parental and crossover gametes. What you will have is the following gametes from the females (left) and males (right):

$bw^+ st^+$	1/4	$bw^+ st^+$	1/2
$bw^+ st$	1/4	$bw\ st$	1/2
$bw\ st^+$	1/4		
$bw\ st$	1/4		

Combining these gametes will give the ratio presented in the table of results.

37. (a) There are several ways to think through this problem. Remember that there is no crossing over in *Drosophila* males. Therefore any gene on the same chromosome will be completely linked to any other gene on the same chromosome. Since you can get *pink* by itself, *short* can not be completely linked to it. This leaves linkage to *black* on the second chromosome, the 4th chromosome, or the X chromosome. Since the distribution of phenotypes in males and females is essentially the same, the gene can not be X-linked. In addition, the F_1 males were wild and if the *short* gene is on the X, the F_1 males would be short.

It is also reasonable to state that the gene can not be on the 4th chromosome because there would be eight phenotypic classes (independent assortment of three genes) instead of the four observed. Through these insights, one could conclude that the *short* gene is on chromosome 2 with the *black* gene.

Another way to approach this problem is to make three chromosomal configurations possible in the F_1 male. By producing gametes from this male, the answer becomes obvious.

Case A		Case B			Case C		
$p\ b$	sh	$p\ sh$	b		$b\ sh$	p	
+ +	+	+	+	+	+ +	+	

Develop the gametes from <u>Case C</u> and cross them out to the completely recessive triple mutant. You will get the results in the table.

(b) The parental cross is now the following:

Females:	$b\ sh\ \ p$	X	Males:	$b\ sh\ \ p$
	+ +\ \ +			$b\ sh\ \ p$

The new gametes resulting from crossing over in the female would be $b +$ and $+ sh$. Since the gene p is assorting independently, it is not important in this discussion. Because 15% of the offspring now contain these recombinant chromatids, the map distance between the two genes must be 15.

152

38. Notice that in the description of the genotype of the female, no mention is made of the *cis-trans* (coupling-repulsion) arrangement of the genes. The data will supply that information. Begin with a set of symbols as indicated below:

B^+ = wild eye shape
B = Bar eye shape

m^+ = wild wings
m = miniature wings

e^+ = wild body color
e = ebony body color

Superficially, the cross would be as follows:

B^+B m^+m e^+e X $B^+?$ $m?$ $e?$ (The ? is used at this point to indicate that we have no information allowing us to decide whether any of the alleles in the male are X-linked.)

Notice from the data that there are approximately as many ebony offspring (282) as those with wild body color (283). Therefore, we can conclude that the *ebony* locus is not linked to B or m. Notice also that the most frequent offspring regarding eye shape and wing size are wild-miniature and Bar-wild. This suggests that the arrangement is "trans" or "repulsion" as indicated below:

B m^+/B^+m; e^+/e

Notice that a semicolon is used to indicate that the *ebony* locus is on a different chromosome.

At this point and without prior knowledge, we still don't know whether any of the genes are X-linked, however, it is of no consequence to the solution of the problem. (In actuality, both B and m loci are X-linked).

To determine the map distances (again, *ebony* is out of the mapping picture at this point because it is not linked to either B or m):

111 + 115 = 226	= wild miniature = parental
117 + 101 = 218	= Bar wild = parental
26 + 31 = 57	= Bar miniature = crossover
29 + 35 = 64	= wild wild = crossover

Mapping the distance between B and m would be as follows:

(57 + 64)/ (226 + 218 + 57 + 64) X 100 =

121/565 X 100 = 21.4 map units.

We would conclude that the *ebony* locus is either far away from B and m (50 map units or more) or it is on a different chromosome. In fact, *ebony* is on a different chromosome.

Chapter 13: Chromosome Mutations: Variation in Chromosome Number and Arrangement

Concept Areas	Corresponding Problems
Variation in Chromosome Number	1, 2, 3, 4, 5, 6, 7, 8, 15, 16, 17, 19, 20, 21, 22, 26, 27
Deletions	9
Duplications	9, 12, 13
Inversions	10, 11, 14, 18, 23, 24, 25
Translocations	14, 28

Vocabulary; Organization and Listing of Terms and Concepts

Structures and Substances

Colchicine

Protoplast

rDNA

Nucleolar organizer (NOR)

Micronucleoli

Processes/ Methods

Chromosome mutations (aberrations)

　aneuploidy(F8.1)

　nondisjunction

　　monosomy, trisomy, tetrasomy,

　　pentasomy

　　Klinefelter syndrome

　　Turner syndrome

　　haplo-VI (*Drosophila*)

partial monosomy

　segmental deletions

　cri-du-chat syndrome, 46,5p-

trisomy

　XXX (*Drosophila*, humans)

　Datura

　pairing configurations

　　trivalent

　Down syndrome (G group)

　　trisomy 21 (47, +21)

　　amniocentesis

　　chorionic villus sampling (CVS)

　　familial Down syndrome

154

Chapter 13 Chromosome Mutations: Variation in Chromosome Number and Arrangement

Patau syndrome (D group)

 trisomy 13 (47, +13)

Edwards syndrome (E group)

 trisomy 18 (47, +18)

 reduced viability

  gametes

  embryos

  spontaneously aborted fetuses

euploidy (F8.1)

 monoploid (n)

 diploid ($2n$)

 polyploid

  triploid ($3n$)

  tetraploid ($4n$)

  pentaploid ($5n$)

 autopolyploidy

  autotriploids ($3n$)

  complete nondisjunction

  dispermic fertilization

   tetraploid X diploid

  autotetraploids ($4n$)

  cold or heat shock

allopolyploidy

 hybridization

allotetraploid

(amphidiploid)

(cotton, *Triticale*)

Raphanus X *Brassica*

 somatic cell hybrids

 (protoplasts)

endopolyploidy

 endomitosis

Chromosome structure

 deletions (deficiency)

  terminal, intercalary

  loop (deficiency, compensation)

  pseudodominance

 duplications

  gene redundancy

   rDNA

    bobbed

  gene amplification

  nucleolar organizer (NOR)

   micronucleoli

  Bar eye in *Drosophila*

   semidominant

   position effect

Chapter 13 Chromosome Mutations: Variation in Chromosome Number and Arrangement

evolutionary aspects

    gene families

    rearrangements

inversions

    paracentric

    pericentric

        arm ratio

    heterozygotes

    inversion loops

    dicentric chromatids

    acentric chromatids

        dicentric bridges

    "suppression of crossing over"

        position effect

        evolutionary aspects

translocations

    reciprocal

    unorthodox synapsis

    semisterility

    familial Down syndrome

    centric fusion

Robertsonian fusion

    14/21 or D/G

fragile sites

    X chromosome

        Martin-Bell syndrome (MBS)

    genetic anticipation

    cancer

Concepts

Significance of variation in chromosomes

    genomic balance

    sex chromosome balance

    evolution

Gene duplication (evolutionary aspects)

    sequence homology

        nucleic acids

        amino acids

Inversions

    "suppression of crossing over"

    evolutionary consequences

Translocations

Fragile sites

Chapter 13 Chromosome Mutations: Variation in Chromosome Number and Arrangement

F13.1 Illustration of the chromosomal configurations of diploid and euploid genomes of *Drosophila melanogaster*.

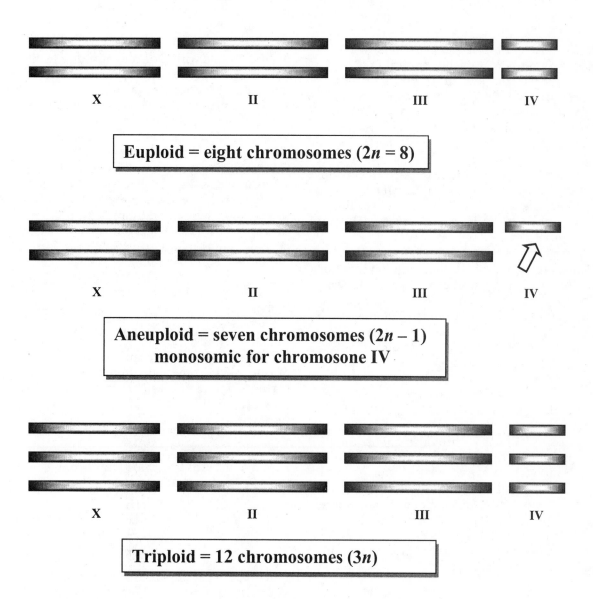

Drosophila melanogaster female

Euploid = eight chromosomes (2*n* = 8)

Aneuploid = seven chromosomes (2*n* – 1) monosomic for chromosone IV

Triploid = 12 chromosomes (3*n*)

Chapter 13 Chromosome Mutations: Variation in Chromosome Number and Arrangement

Solutions to Problems and Discussion Questions

1. With a diploid chromosome number of 18 (2n), a haploid (n) would have nine chromosomes, a triploid (3n) would have 27 chromosomes, and a tetraploid (4n) would have 36 chromosomes. A trisomic would have one extra chromosome (19) and a monosomic one less than the diploid (17).

2. With frequent exceptions especially in plants, organisms typically inherit one chromosome complement (*haploid* = n = one representative of each chromosome) from each parent. Such organisms are *diploid*, or 2n. When an organism contains complete multiples of the n complement (3n, 4n, 5n, etc.) it is said to be *euploid* in contrast to aneuploid in which complete haploid sets do not occur. An example of an aneuploid is *trisomic* where a chromosome is added to the 2n complement. In humans, trisomy 21 would be symbolized as 2n+1 or 47,+21.

Monosomy is an aneuploid condition in which one member of a chromosome pair is missing, thus producing the chromosomal formula of 2n-1. Haplo-IV is an example of monosomy in *Drosophila*. *Trisomy* is the chromosomal condition of 2n+1 where an extra chromosome is present. Down syndrome is an example in humans (47, +21). See the K/C text and notice that all the chromosomes are present in the diploid state except chromosome #21.

Patau syndrome is a chromosomal condition where there is an extra D group chromosome. Such individuals are 47,+13 and have multiple congenital malformations. *Edwards syndrome* is a chromosomal condition where there is an extra E group chromosome (47,+18). Individuals with Edwards syndrome have multiple congenital malformations and reduced life expectancy.

Polyploidy refers to instances where there are more than two haploid sets of chromosomes in an individual cell. *Autopolyploidy* refers to cases of polyploidy where the chromosomes in the individual originate from the same species.

Allopolyploidy involves instances where the chromosomes originate from the hybridization of two different species, usually closely related.

3. Individuals with Down syndrome, while suffering congenital defects, tendencies toward respiratory disease and leukemia, can live well into adulthood. Individuals with Patau or Edwards syndrome live less than four months on the average. Comparing the different sizes of the involved chromosomes (21, 13, and 18, respectively) in the K/C text for example, suggests that the larger the chromosome, the lower the likelihood of lengthy survival. In addition, it would be expected that certain chromosomes, because of their genetic content, may have different influences on development.

4. The fact that there is a significant maternal age effect associated with Down syndrome indicates that nondisjunction in older females contributes disproportionately to the number of Down syndrome individuals. In addition, certain genetic and cytogenetic marker data indicate the influence of female nondisjunction.

5. While several trisomies (for chromosomes 21, 18, 13, the X and Y) are tolerated, monosomy for the autosomes is not tolerated. Karyotypic analysis of spontaneously aborted fetuses has indicated a relatively large degree of departures from the typical diploid state. The delicate genetic balance produced by millions of years of evolution must be maintained in order for any organism (but especially animals) to develop normally.

Monosomy leads to the exposure of recessive, deleterious genes thus producing developmental abnormalities. Dosage compensation of the sex chromosomes and the relative paucity of Y-linked genes probably contribute to the survival of sex-chromosome aneuploidy. Notice how large the X chromosome is compared with other chromosomes.

At least 20 percent of all conceptions are terminated in natural abortion. Of these, thirty percent show some chromosomal anomaly. Of the chromosomal anomalies that occur, approximately ninety percent are eliminated by spontaneous abortion.

Trisomy for every human chromosome has been observed, however, monosomy, the reciprocal meiotic event of trisomy, is rare. This observation probably results from gamete or early embryonic inviability.

6. Because an allotetraploid has a possibility of producing bivalents at meiosis I, it would be considered the most fertile of the three. Having an even number of chromosomes to match up at the metaphase I plate, autotetraploids would be considered to be more fertile than autotriploids.

7. The sterility of interspecific hybrids is often caused from a high proportion of univalents in meiosis I. As such, viable gametes are rare and the likelihood of two such gametes "meeting" is remote. Even if partial homology of chromosomes allows some pairing, sterility is usually the rule. The horticulturist may attempt to reverse the sterility by treating the sterile hybrid with colchicine. Such a treatment, if sucessful, may double the chromosome number and each chromosome would now have a homologue with which to pair during meiosis.

8. American cultivated cotton has 26 pairs of chromosomes; 13 large, 13 small. Old world cotton has 13 pairs of large chromosomes and American wild cotton has 13 pairs of small chromosomes. It is likely that an interspecific hybridization occurred followed by chromosome doubling. These events probably produced a fertile amphidiploid (allotetraploid). Experiments have been conducted to reconstruct the origin of American cultivated cotton.

9. Basically the synaptic configurations produced by chromosomes bearing a deletion or duplication (on one homologue) are very similar. There will be point-for-point pairing in all sections which are capable of pairing. The section which has no homologue will "loop out" as in the K/C text.

10. While there is the appearance that crossing over is suppressed in inversion "heterozygotes" the phenomenon extends from the fact that the crossover chromatids end up being abnormal in genetic content. As such they fail to produce viable (or competitive) gametes or lead to zygotic or embryonic death. Notice in the K/C text the crossover chromatids end up genetically unbalanced.

11. Examine the K/C text and notice that in (a) there are two genetically balanced chromatids (normal and inverted) and two, those resulting from a single crossover in the inversion loop, which are genetically unbalanced and abnormal (dicentric and acentric). The dicentric chromatid will often break, thereby producing highly abnormal fragments whereas the acentric fragment is often lost in the meiotic process. In part (b) all the chromatids have centromeres, but the two chromatids involved in the crossover are genetically unbalanced. The balanced chromatids are of normal or inverted sequence.

12. The mutant *Notch* in *Drosophila* produces flies with abnormal wings. It is a sex-linked dominant gene (a deletion) which also behaves as a recessive lethal. A deficiency (compensation) loop indicates that bands 3C2 through 3C11 are involved. Loci near *Notch* display pseudodominance. The *Bar* gene on the other hand results from a duplication of a sex-linked region (16A) and results in abnormal eye shape.

Females: N^+/N	X	males: B/Y
1/4	N^+/B	females; Bar
1/4	N/B	females; Notch, Bar
1/4	N^+/Y	males; wild
1/4	N/Y	**lethal**

The final phenotypic ratio would be 1:1:1 for the phenotypes shown above.

If a gene exists in a duplicated state and if that gene's product is required for survival, then mutation in either (but not both) the original gene or its duplicate will not ordinarily threaten the survival of the organism. Duplication of a gene provides a buffer to mutation.

13. In a work entitled *Evolution by Gene Duplication*, Ohno, suggests that gene duplication has been essential in the origin of new genes. If gene products serve essential functions, mutation and therefore evolution, would not be possible unless these gene products could be compensated for by products of duplicated, normal genes. The duplicated genes, or the original genes themselves, would be able to undergo mutational "experimentation" without necessarily threatening the survival of the ogranism.

14. It is likely that when certain combinations of genes are of selective advantage in a specific and stable environment, it would be beneficial to the organism to protect that gene combination from disruption through crossing over. By having the genes in an inversion, crossover chromatids are not recovered and therefore are not passed on to future generations.

Translocations offer an opportunity for new gene combinations by associations of genes from nonhomologous chromosomes. Under certain conditions such new combinations may be of selective advantage and meiotic conditions have evolved so that segregation of translocated chromosomes yields a relatively uniform set of gametes.

15. A Turner syndrome female has the sex chromosome composition of XO. If the father had hemophilia it is likely that the Turner syndrome individual inherited the X chromosome from the father and no sex chromosome from the mother. If nondisjunction occurred in the mother, either during meiosis I or meiosis II, an egg with no X chromosome can be the result. See the K/C text for a diagram of primary and secondary nondisjunction.

16. The primrose, *Primula kewensis*, with its 36 chromosomes, is likely to have formed from the hybridization and subsequent chromosome doubling of a cross between the two other species, each with 18 chromosomes. An example of this type of allotetraploidy (amphidiploidy) is seen in the K/C text.

17. Given the basic chromosome set of nine unique chromosomes (a haploid complement) other forms with the "*n* multiples" are forms of autotetraploidy. In the illustration below the *n* basic set is multiplied to various levels as is the autotetraploid in the example.

Basic Set of nine unique chromosomes (n)

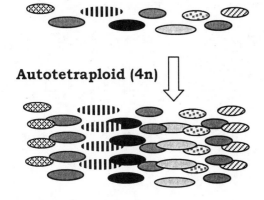

Autotetraploid (4n)

Individual organisms with 27 chromosomes (3*n*) are more likely to be sterile because there are trivalents at meiosis I which cause a relatively high number of unbalanced gametes to be formed.

18. The rare double crossovers in the boundaries of a paracentric or pericentric inversion produce only minor departures from the standard chromosomal arrangement as long as the crossovers involve the same two chromatids. With two-strand double crossovers, the second crossover negates the first. However, three-strand and four-strand double crossovers have consequences which lead to anaphase bridges as well as a high degree of genetically unbalanced gametes.

19. Set up the cross in the usual manner, realizing that recessive genes in the Haplo-IV individual will be expressed.

Let *b* = bent bristles; *b⁺* = normal bristles

(a)

_/ *b* X *b⁺* / *b⁺*

F₁:

| _/ *b⁺* | = normal bristles |
| *b* / *b⁺* | = normal bristles |

F₂:

_/ *b ⁺* X *b* / *b⁺*

_/ *b⁺*	= normal bristles
_/ *b*	= bent bristles
b ⁺/ *b⁺*	= normal bristles
b/ *b⁺*	= normal bristles

(b)

_/ *b ⁺* X *b*/*b*

F₁:

| _/ *b* | = bent bristles |
| *b* / *b⁺* | = normal bristles |

F₂:

_/ *b* X *b* / *b⁺*

_/ *b⁺*	= normal bristles
_/ *b*	= bent bristles
b ⁺/ *b*	= normal bristles
b / *b*	= bent bristles

20. In the trisomic, segregation will be "2 X 1" as illustrated below:

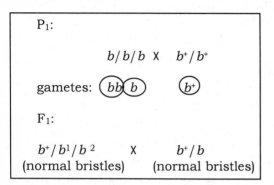

P₁:

$b/b/b$ X b^+/b^+

gametes: (*bb*)(*b*)   (*b⁺*)

F₁:

$b^+/b^1/b^2$ X b^+/b
(normal bristles)   (normal bristles)

Notice that there are several segregation patterns created by the trivalent at anaphase I.

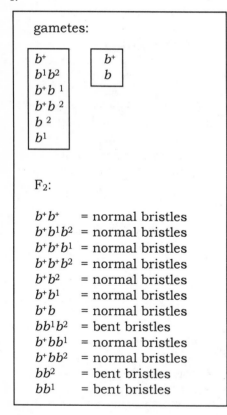

gametes:

b^+	b^+
b^1b^2	b
b^+b^1	
b^+b^2	
b^2	
b^1	

F₂:

b^+b^+	= normal bristles
$b^+b^1b^2$	= normal bristles
$b^+b^+b^1$	= normal bristles
$b^+b^+b^2$	= normal bristles
b^+b^2	= normal bristles
b^+b^1	= normal bristles
b^+b	= normal bristles
bb^1b^2	= bent bristles
b^+bb^1	= normal bristles
b^+bb^2	= normal bristles
bb^2	= bent bristles
bb^1	= bent bristles

21. The cross would be as follows:

$$WWWW \quad X \quad wwww$$

(assuming that chromosomes pair as bivalents at meiosis)

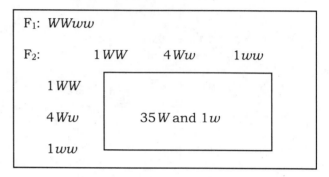

F₁: *WWww*

F₂: 1 *WW*   4 *Ww*   1 *ww*

1 *WW*

4 *Ww*   35 *W* and 1 *w*

1 *ww*

22. Given some of the information in the above problem the expression would be as follows:

$(35/36\,W : 1/36\,w)(35/36\,A : 1/36\,a)$ ⟹

(35/36)² $W\text{---}A\text{---}$

35/(36)² $W\text{---}aaaa$

35/(36)² $wwwwA\text{---}$

1/(36)² $wwwwaaaa$

23. (a) In all probability, crossing over in the inversion loop of an inversion (in the heterozygous state) had produced defective, unbalanced chromatids, thus leading to stillbirths and/or malformed children. **(b)** It is probable that a significant proportion (perhaps 50%) of the children of the man will be similarly influenced by the inversion. **(c)** Since the karyotypic abnormality is observable, it may be possible to detect some of the abnormal chromosomes of the fetus by amniocentesis or CVS. However, depending on the type of inversion and the ability to detect minor changes in banding patterns, not all abnormal chromosomes may be detected.

24. Considering that there are at least three map units between each of the loci, and that only four phenotypes are observed, it is likely that genes *a b c d* are included in an inversion and crossovers which do occur among these genes are not recovered because of their genetically unbalanced nature. In a sense, the minimum distance between loci *d* and *e* can be estimated as 10 map units

$$(48 + 52/1000);$$

However, this is actually the distance from the *e* locus to the breakpoint which includes the inversion.

The "map" is therefore as drawn below:

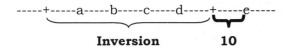

25. (a) Reciprocal translocation

(b)

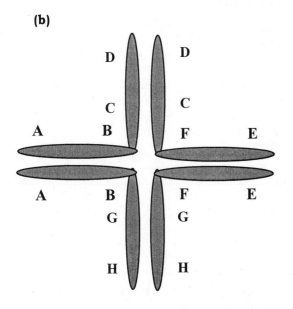

(c) Notice that all chromosomal segments are present and there is no apparent loss of chromosomal material. However, if the breakpoints for the translocation occurred within genes then an abnormal phenotype may be the result. In addition, a gene's function is sometimes influenced by its position; its neighbors in other words. If such "position effects" occur then a different phenotype may result.

(d) It is likely that the translocation is the cause of the miscarriages. Segregation of the chromosomal elements will produce approximately half unbalanced gametes. The chance of a normal child is approximately one in two, however half of the normal children will be translocation carriers.

26. (a) The father must have contributed the abnormal X-linked gene.

(b) Since the son is XXY and heterozygous for anhidrotic dysplasia, he must have received both the defective gene and the Y chromosome from his father. Thus non-disjunction must have occurred during meiosis I.

(c) This son's mosaic phenotype is caused by X-chromosome inactivation, a form of dosage compensation in mammals.

27. Notice that a chromosome in this question is defined as having two sisters joined at the centromere. This is the expected chromosome structure at the end of meiosis I.

(a) In light of this information, meiosis I must have produced the abnormal oocytes with more or less than 24 chromosomes, indicating multiple conditions of nondisjunction. More likely, the oocytes consisted of "$22_{1/2}$" chromosomes, those 22 normal dyads and a single monad.

(b) The result will be a monosomic and a normal zygote assuming that the half chromosome (monad) migrates, intact, to one pole or the other.

(c) In all likelihood, premature division of the centromere (at meiosis I) probably causes the single (non-duplicated) chromosome at meiosis II.

(d) We generally consider nondisjunction occurring at meiosis I to consist of intact chromosomes, two sister chromatids, failing to separate appropriately. These data indicate that some forms of aneuploidy result from premature division of the centromere at meiosis I as in the figure below.

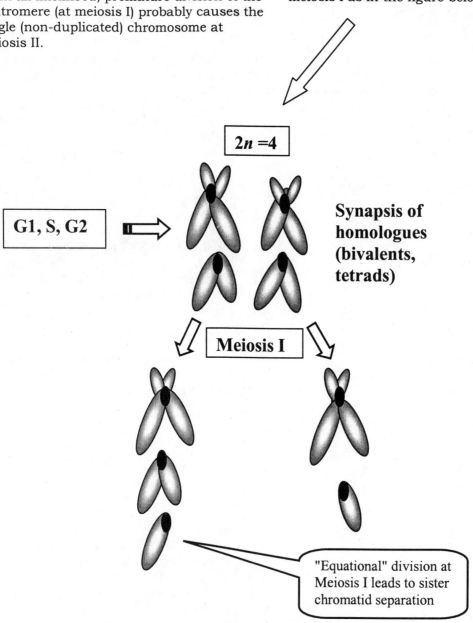

2n = 4

G1, S, G2

Synapsis of homologues (bivalents, tetrads)

Meiosis I

"Equational" division at Meiosis I leads to sister chromatid separation

Chapter 13 Chromosome Mutations: Variation in Chromosome Number and Arrangement

28. First consider what is meant be a Robertsonian translocation: breaks at the short arms of two nonhomologous acrocentric chromosomes where the small acentric fragments are lost and the larger chromosomal segments fuse at or near the centromeric region, producing a compound, larger submetacentric or metacentric chromosome. Below is a description of breakage/reunion events which illustrate such a translocation in relatively small, similarly sized, chromosomes 19 (metacentric) and 20(metacentric/submetacentric). The case described here is shown occurring before S phase duplication. The same phenomenon is shown in the text as occurring after S phase. Since the likelihood of such a translocation is fairly small in a general population, inbreeding played a significant role in allowing the translocation to "meet itself."

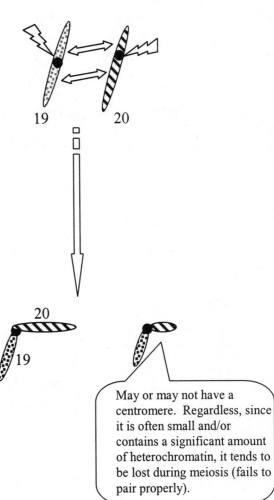

After S phase

May or may not have a centromere. Regardless, since it is often small and/or contains a significant amount of heterochromatin, it tends to be lost during meiosis (fails to pair properly).

Chapter 14: Extranuclear Inheritance

Concept Areas	Corresponding Problems
Extrachromosomal Inheritance	1, 13, 16
Maternal Effect	9, 10, 11, 12, 13, 14, 15
Organelle Heredity	2, 3, 4, 5, 6
Infectious Heredity	7, 8

Vocabulary: Organization and Listing of Terms and Concepts

Structures and Substances

Chloroplast DNA

Mitochondrial DNA

Heterokaryon

Conidia

Kappa

Paramecin

Sigma

Kynurenine

Tryptophan

Processes/Methods

Organelle heredity
(cytoplasmic inheritance)

   chloroplast DNA (cpDNA)

      coding products

         rRNA, tRNA

         ribosomal proteins, *etc.*

   Mirabilis jalapa

RuBP
(ribulose-1-5-bisphosphate carboxylase)

Chlamydomonas reinhardi

mt^+, mt^-

mitochondrial DNA (mtDNA)

coding products

rRNA, tRNA, proteins

Neurospora crassa (*poky*)

mi-1

suppressive mutations

Saccharomyces cerevisiae (*petite*)

segregational

neutral

suppressive

endosymbiotic hypothesis

Chapter 14 Extranuclear Inheritance

humans

heteroplasmy

myoclonic epilepsy (MERRF)

Leber's hereditary optic

neuropathy (LHON)

Kearns-Sayre syndrome

Infectious heredity

Paramecium aurelia

killers, paramecin

kappa

conjugation

autogamy

Drosophila

CO_2 sensitivity

D. bifasciata, D. willistoni

sex ratio

maternal effect (influence)

Ephestia kuhniella (*A, a*)

Limnaea peregra (*D, d*)

dextral, sinistral

first cleavage division

spindle orientation

injection experiments

molecular gradients

bicoid

Genomic imprinting

dosage compensation

Prader-Willi syndrome (PWS)

Angelman syndrome (AS)

DNA methylation

Concepts

Extranuclear inheritance (F14.1)

Products of cpDNA and mtDNA

Non-Mendelian patterns of inheritance (F14.1)

maternal, infectious, organelle

Genomic imprinting

F14.1 Illustration of the common pattern seen in many cases of extranuclear inheritance. The condition of the female (egg) parent has a stronger influence on the phenotype of the offspring than the male (sperm/pollen) parent. Reciprocal crosses give different results in offspring.

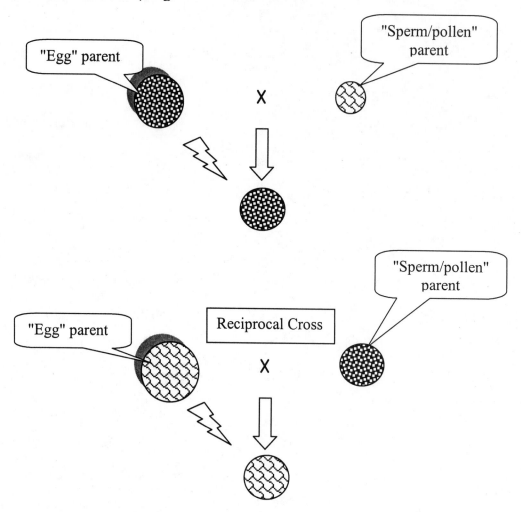

Chapter 14 Extranuclear Inheritance

Solutions to Problems and Discussion Questions

1. In cases of extrachromosomal inheritance, the phenotype is determined by the nuclear (maternal effect) or cytoplasmic (organelle or infectious) condition of the parent which contributes the bulk of the cytoplasm to the offspring. In most cases, the maternal parent provides the basis for the cytoplasmic inheritance.

The pattern of inheritance is more often from one parent to the offspring. One does not see both parents contributing to the characteristics of the offspring as is the case with Mendelian (chromosomal) forms of inheritance. Standard Mendelian ratios (3:1) are usually not present. In general, the results of reciprocal crosses differ. See F14.1.

```
Female mutant   X    male wild

        all offspring mutant
   _____

Female wild   X    male mutant

        all offspring wild
```

In sex-linked inheritance, the pattern is often from grandfather through carrier mother to son. Patterns of extrachromosomal inheritance are often not influenced by the sex of the individual.

2. The *mt⁺* strain (resistant for the nuclear and chloroplast genes) contributes the "cytoplasmic" component of streptomycin resistance which would negate any contribution from the *mt⁻* strain. Therefore, all the offspring will have the streptomycin resistance phenotype. In the reciprocal cross, with the *mt⁺* strain being streptomycin sensitive, all the offspring will be sensitive.

3. Because the ovule source furnishes the cytoplasm to the embryo and thus the chloroplasts, the offspring will have the same phenotype as the plant providing the ovule.

a) green

b) white

c) white, variegated, or green

d) green

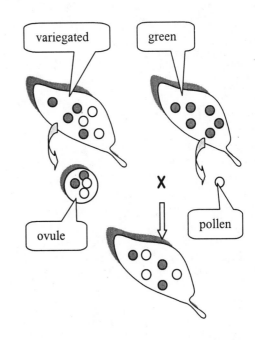

4. See the K/C text for a comparison of results involving various petite strains.

(a) neutral
(b) segregational (nuclear mutations)
(c) suppressive

5. As with any description of dominance, one looks to the phenotype of the diploid heterozygote. In this problem, the heterozygote is of normal phenotype, therefore the *petite* gene is recessive.

6. Examine the K/C text and notice that the inheritance patterns for the two, *segregational* and *neutral*, are quite different. The segregational mode is dependent on nuclear genes while that of the neutral type is dependent on cytoplasmic influences, namely mitochondria. If the two are crossed as stated in the problem, then one would expect, in the diploid zygote, the *segregational* allele to be "covered" by normal alleles from the neutral strain. On the other hand, as the nuclear genes are again "exposed" in the haploid state of the ascospores, one would expect a 1:1 ratio of normals to petites. The petite phenoytpe is caused by the nuclear, segregational gene.

7. In providing the answers to this question, remember that a *Paramecium* may be sensitive and carry the K gene. These are organisms which did not obtain kappa particles. There is a question as to whether cytoplasmic exchange has occurred; however, seeing the results in cross (b) one can assume that cytoplasmic exchange has occurred. In addition, one can assume that only the exconjugants are being described in the offspring. Under those conditions, the parental genotypes could be the following:

(a) *Kk* X *kk*

(b) any case where there is no *kk* such as:

     KK   X   *KK*
           or
     KK   X   *kk*

(c)   *Kk*   X   *Kk*

8. (a) There are many similarities among mitochondrial, chloroplast, and prokaryotic molecular systems. It is likely that mitochondria and chloroplasts evolved from bacteria in a symbiotic relationship, therefore it is not surprising that certain antibiotics which influence bacteria will also influence all mitochondria and chloroplasts.

(b) Clearly, the *mt*+ strain is the donor of the *cp*DNA since the inheritance of resistance or sensitivity is dependent on the status of the *mt*+ gene.

9. The case with *Limnaea* involves a maternal effect in which the *genotype* of the mother influences the *phenotype* of the *immediate* offspring in a non-Mendelian manner. Notice that in the above statement, it is the maternal genotype which determines the phenoytpe of the offspring, regardless of its own genotype.

Since both of the parents are *Dd*, the parent contributing the eggs must be *Dd*. Therefore, all of the offspring must have the phenotype of the mother's genotype, which is dextral.

10. In a maternal effect, the *genotype* of the mother influences the *phenotype* of her immediate offspring in a non-Mendelian manner. The fact that all of the offspring (F_1) showed a dextral coiling pattern indicates that one of the parents (maternal parent) contains the *D* allele. Taking these offspring and seeing that their progeny (call these F_2) occur in a 1:1 ratio indicates that half of the offspring (F_1) are *dd*. In order to have these results, one of the original parents must have been *Dd* while the other must have been *dd*.

Parents: *Dd*   X   *dd*

Offspring (F1):   1/2 *Dd*,   1/2 *dd*

(all dextral because of the maternal genotype)

Progeny (F_2):

All those from *Dd* parents will be dextral while all those from *dd* parents will be sinistral.

---

11. It appears as if some factor normally provided by the gs^+ allele is necessary for normal development and/or functioning of the female offspring's gonads. Without this product, the daughters are sterile, thus the term "grandchildless." Because the female provides so much vital material and information to the egg, including the cytoplasm necessary for germ line determination, it is not surprising that such maternal effect genes exist.

12. Since there is no evidence for segregation patterns typical of chromosomal genes and Mendelian traits, some form of extranuclear inheritance seems possible. If the *lethargic* gene is dominant then a maternal effect may be involved. In that case, some of the F_2 progeny would be hyperactive because maternal effects are only temporary, affecting only the immediate progeny. If the lethargic condition is caused by some infective agent, then perhaps injection experiments could be used. If caused by a mitochondrial defect, then the condition would persist in all offspring of lethargic mothers, through more than one generation.

13. Developmental phenomena which occur early are more likely to be under maternal influence than those occurring late. Anterior/posterior and dorsal/ventral orientations are among the earliest to be established and in organisms where their study is experimentally and/or genetically approachable, they often show considerable maternal influence. Maternal effect genes produce products which are not carried over for more than one generation as is the case with organelle and infectious heredity. Crosses which illustrate the transient nature of a maternal effect could include the following. However, depending on particular biochemical/developmental parameters, all crosses may not give these types of patterns.

Female *Aa* X male *aa* -----> all offspring of the "A" phenotype. Take a female "A" phenotype from the above cross and conduct the following mating: *aa* X male *Aa* ----->.

All offspring may be of the "a" phenotype because all of the offspring will reflect the *genotype* of the mother, not her *phenotype*. This cross illustrates that maternal effects last only one generation. In actual practice, the results of this cross may give a typical 1:1 ratio, depending on the biochemical/developmental characteristics of the system. However, the fact that the maternal effect only persists for one generation is clearly illustrated regardless of which set of results occurs.

14. (a) The presence of bcd^-/bcd^- males can be explained by the maternal effect: mothers were bcd^+/bcd^-. **(b)** The cross

$$\text{female } bcd^+/bcd^- \text{ X male } bcd^-/bcd^-$$

will produce an F_1 with normal embryogenesis because of the maternal effect. In the F_2, any cross having bcd^+/bcd^- mothers will have phenotypically normal embryos. Any cross involving homozygous bcd^-/bcd^- mothers will have problems with embryogenesis.

15. Because of sampling error due to relatively small numbers of offspring, this pedigree could represent a typical Mendelian dominant or recessive gene; however, because the thrust of this chapter is on extranuclear inheritance, the condition is probably a case of extranuclear inheritance. The phenotype of the offspring will therefore follow the characteristics of the mother.

16. (a) A locus, *Segregation Distortion* (*SD*), is present on the wild type chromosome. Some aspect of *SD* causes a shift in the segregation ratio by allowing sperm to carry the *SD* chromosome at the expense of the homologue. **(b)** One could use this *SD* chromosome in a variety of crosses and determine that the abnormal segregation is based on a particular chromosomal element. One could even map the *SD* locus on the second chromosome (as has been done). **(c)** Segregation Distortion describes a condition in which typical Mendelian segregation is distorted from the 50:50.

Chapter 15: Genetics of Bacteria and Bacteriophages

Concept Areas	Corresponding Problems
Bacterial Mutation and Growth	
Genetic Recombination in Bacteria	1
Conjugation	2, 3, 4, 5, 6, 19, 22
Transformation	7, 8, 9, 21
Bacteriophages	15, 16
Transduction	10, 11, 12, 13
Mutation and Recombination in Viruses	14
Intragenic Recombination in Phage	17, 18, 20

Vocabulary: Organization and Listing of Terms and Concepts

Structures and Substances

Bacteria

Bacteriophage

Spontaneous mutations

Growth conditions

   minimal medium

   liquid culture

   prototroph

   auxotroph

Donor strain

   E. coli K12

   F sex pilus

   fertility factor, F factor

   RecA, RecBCD proteins

   rec genes

lysozyme

Hfr, circular chromosome

F', merozygotes

   partial diploid

Plasmids

   F factors

   R plasmids

     resistance transfer factor (RTF)

     r-determinants

     antibiotic resistance

   Col plasmids

     ColE1

     colicins

     colicinogenic

172

Chapter 15 Genetics of Bacteria and Bacteriophages

Protein capsid

Lysozyme

Plaque

Episome

Prophage P22

Cistron

Hot spot

Processes/Methods

Sensitive

Resistant

  lag, log, stationary phases

Bacterial recombination

  conjugation

    F+, F-

    physical contact

    unidirectional

    donor, "male"

    recipient, "female"

    high frequency recombination, Hfr

      oriented transfer

      interrupted mating technique

      circular map

  F' state

    merozygotes

Transformation

  competence

  heteroduplex

  linkage

  cotransformation

Transduction

  phage life cycle

    plaque (plaque assay)

      lysis

    lysogeny

      temperate phage

      symbiotic relationship

      prophage

      lysogenic bacterium

      U-tube experiment

        filterable agent (FA)

        prophage P22

    generalized transduction (F15.1)

      abortive transduction

      complete transduction

    cotransduction

    mapping

      specialized transduction

        prophage λ

          att, gal, bio

Mutations (viral)

   rapid lysis, host range

   mixed infection experiments

   negative interference

     intragenic exchanges

      fine structure analysis

        *r*II, T4

        E. coli B, K12

    complementation

    deletion testing

      hot spots

Concepts

   Dilution

   Bacterial recombination - all forms

     relationship to *rec* genes

   Fine structure analysis

     complementation

     cistron

     recombinational analysis

     deletion testing

_____

F15.1 Simple illustration comparing *abortive* and *complete* transduction.

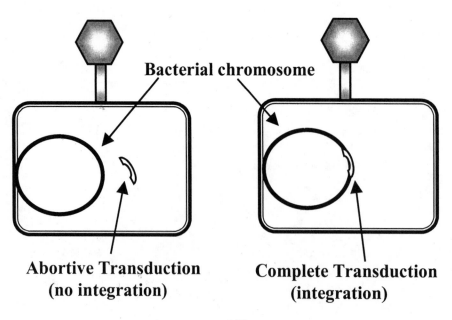

Abortive Transduction
(no integration)

Complete Transduction
(integration)

Solutions to Problems and Discussion Questions

1. Three modes of recombination in bacteria are *conjugation, transformation,* and *transduction.* Conjugation is dependent on the F factor which, by a variety of mechanisms, can direct genetic exchange between two bacterial cells. Transformation is the uptake of exogenous DNA by cells. Transduction is the exchange of genetic material using a bacteriophage.

2. (a) The requirement for physical contact between bacterial cells during conjugation was established by placing a filter in a U-tube such that the medium can be exchanged but the bacteria can not come in contact. Under this condition, conjugation does not occur.

(b) By treating cells with streptomycin, an antibiotic, it was shown that recombination would not occur if one of the two bacterial strains was inactivated. However, if the other was similarly treated, recombination would occur. Thus, directionality was suggested, with one strain being a donor strain and the other being the recipient.

(c) An F⁺ bacterium contains a circular, double-stranded, structurally independent, DNA molecule which can direct recombination.

3. (a) In an F⁺ X F⁻ cross, the transfer of the F factor produces a recipient bacterium which is F⁺. Any gene may be transferred, and the frequency of transfer is relatively low. Crosses which are Hfr X F⁻ produce recombinants at a higher frequency than the F⁺ X F⁻ cross. The transfer is oriented (non-random) and the recipient cell remains F⁻.

(b) Bacteria which are F⁺ possess the F factor, while those that are F⁻ lack the F factor. In Hfr cells the F factor is integrated into the bacterial chromosome and in F' bacteria, the F factor is free of the bacterial chromosome yet possesses a piece of the bacterial chromosome.

4. Mapping the chromosome in an Hfr X F⁻ cross takes advantage of the oriented transfer of the bacterial chromosome through the conjugation tube. For each F type, the point of insertion and the direction of transfer are fixed, therefore breaking the conjugation tube at different times produces partial diploids with corresponding portions of the donor chromosome being transferred. The length of the chromosome being transferred is contingent on the duration of conjugation, thus mapping of genes is based on time.

5. In an Hfr X F⁻ cross, the F factor is directing the transfer of the donor chromosome. It takes approximately 90 minutes to transfer the entire chromosome. Because the F factor is the last element to be transferred and the conjugation tube is fragile, the likelihood for complete transfer is low.

6. As shown in the K/C text, the F⁺ element can enter the host bacterial chromosome and upon returning to its independent state, it may pick up a piece of a bacterial chromosome. When combined with a bacterium with a complete chromosome, a partial diploid, or merozygote, is formed.

7. Transformation requires *competence* on the part of the recipient bacterium, meaning that only under certain conditions are bacterial cells capable of being transformed. Transforming DNA must be *double-stranded* to begin with yet is converted to a single-stranded structure upon insertion into the host cell. The most efficient length of the transforming DNA is about 1/200 of the size of the host chromosome. Transformation is an energy-requiring process and the number of sites on the bacterial cell surface is limited.

8. In the first data set, the transformation of each locus, a^+ or b^+, occurs at a frequency of .031 and .012 respectively. To determine if there is linkage one would determine whether the frequency of double transformants a^+b^+ is greater than that expected by a multiplication of the two independent events. Multiplying .031 X .012 gives .00037 or approximately 0.04%. From this information, one would consider no linkage between these two loci. Notice that this frequency is approximately the same as the frequency in the second experiment, where the loci are transformed independently.

9. Notice that the incorporation of loci a^+ and b^+ occurs much more frequently than the incorporation of b^+ and c^+ together (210 to 1) and the incorporation of all three genes $a^+b^+c^+$ occurs relatively infrequently. If a and b loci are close together and both are far from locus c then fewer crossovers would be required to incorporate the two linked loci compared to all three loci (see diagram). If all three loci were close together then the frequency of incorporation of all three would be similar to the frequency of incorporation of any two contiguous loci, which is not the case.

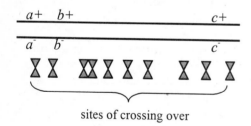

sites of crossing over

10. In their experiment a filter was placed between the two auxotrophic strains which would not allow contact. F-mediated conjugation requires contact and without that contact, such conjugation can not occur. The treatment with DNase showed that the filterable agent was not naked DNA.

11. A *plaque* results when bacteria in a "lawn" are infected by a phage and the progeny of the phage destroy (lyse) the bacteria. A somewhat clear region is produced which is called a plaque.

Lysogeny is a complex process whereby certain temperate phage can enter a bacterial cell and instead of following a lytic developmental path, integrate their DNA into the bacterial chromosome. In doing so, the bacterial cell becomes lysogenic. The latent, integrated phage chromosome is called a *prophage.*

12. In *generalized transduction* virtually any genetic element from a host strain may be included in the phage coat and thereby be transduced. In *specialized (restricted) transduction* only those genetic elements of the host which are closely linked to the insertion point of the phage can be transduced. Specialized transduction involves the process of lysogeny.

Because only certain genetic elements are involved in specialized transduction, it is not useful in determining linkage relationships. Cotransduction of genes in generalized transduction allows linkage relationships to be determined.

13. The first problem to be solved is the gene order. Clearly, the parental types are

$$a^+b^+c^+ \text{ and } a^-b^-c^-$$

because they are the most frequent. The double crossover types are the least frequent,

$$a^-b^-c^+ \text{ and } a^+b^+c^-.$$

Because it is the gene in the middle which switches places when one compares the parental and double crossover classes, the *c* gene must be in the middle. The map distances are as follows:

a to *c* = (740 + 670 + 90 + 110)/10,000

= 16.1 map units

c to *b* = (160 + 140 + 90 + 110)/10,000

= 5 map units

To determine the type of interference, first determine the *expected* frequency of double crossovers (0.161 X .05 = .000805), which when multiplied by 10,000 gives approximately 80. The *observed* number of double crossovers is 90 + 110 or 200. Since many more double crossovers are observed than expected, negative interference is occurring.

14. Viral recombination occurs when there is a sufficiently high number of infecting viruses so that there is a high likelihood that more than one type of phage will infect a given bacterium. Under this condition phage chromosomes can recombine by crossing over.

15. Starting with a single bacteriophage, one lytic cycle produces 200 progeny phage, three more lytic cycles would produce (200)^4 or 1,600,000,000 phage.

16. (a) The concentration of phage is greater than 10^4.

(b) The concentration of phage is around 1.4 X 10^6.

(c) The concentration of phage is less than 10^6.

17. The approach for determining the complementation groupings and the results of the missing data is to recall that if a "+" is registered, different complementation groups (genes) exist. If a "-" results, then the two mutations are in the same complementation group. For Group A, *d* and *f* are in the same complementation group (gene) while *e* is in a different one. Therefore

e X *f* = +.

For Group B, all three mutations are in the same gene, hence

h X *i* = -.

In Group C, *j* and *k* are in different complementation groups as are *j* and *l*. It would be impossible to determine whether *l* and *k* are in the same or different complementation group if the *r*II region had more than two cistrons. However, because only two complementation regions exist, and both are not in the same one as *j*, *k* and *l* must both be in the other.

18. Because there are only two complementation groups in the *r*II region one would have the following groupings:

Group A: 1,4,5   *Group* B: 2,3

(a) Therefore the result of testing

2 X 3 = no lysis;
2 X 4 = lysis;
3 X 4 = lysis.

(b) Because mutant 5 failed to complement with mutations in either cistron, it probably represents a major alteration in the gene such that both cistrons are altered. A deletion which overlaps both cistrons could cause such a major alteration.

(c) The recombination frequency is given by the following formula. Recall that only one of the two recombinant types is recovered in this type of experiment where the assay of growth on *E. coli* B is used. Remember to include the dilution factor in the setting of the observed values.

General formula:

$$\frac{2(\text{number of recombinant types})}{\text{total number of progeny}}$$

$$= 2(5 \times 10^1)/(2 \times 10^5) = 5 \times 10^{-4}$$

(d) Because mutant 6 complemented mutations 2 and 3, it is likely to be in the cistron with mutants 1,4, and 5. A lack of recombinants with mutant 4 indicates that mutant 6 is a deletion which overlaps mutation 4. Recombinants with 1 and 5 indicates that the deletion does not overlap these mutations.

19. One can approach this problem by lining up the data from the various crosses in the following order:

Hfr Strain	Order
1	T C H R O
2	H R O M B
3	<<CH R O M
4	M B A K T>>
5	< <BAKTC

Overall:

T C H R O M B A K

Notice that all of the genes can be linked together to give a consistent map and that the ends overlap, indicating that the map is circular. The order is reversed in two of the crosses indicating the orientation of transfer is reversed.

20. (a)

Combination	Complementation
1, 2	-
1, 3	+
2, 4	+
4, 5	-

(b) Because mutants can lyse *E. coli* B but not K12 one can determine the total number of plaque forming units (phage) as 4×10^7. The number of recombinants (those that grow on K12) would be 8×10^2. The recombination frequency would therefore be

$$2(8 \times 10^2/4 \times 10^7) = 4 \times 10^{-5}$$

(c) The dilution would be 10^{-3} and the colony number would be 8×10^3.

(d) Mutant 7 might well be a deletion spanning parts of both A and B cistrons.

21. Because the frequency of double transformants is quite high (compare the *try⁺tyr⁺* transformants in A and B experiments) one may conclude that the genes are quite closely linked together. Part B in the experiment gives one the frequencies of transformations of the individual genes and the frequency of transformants receiving two pieces of DNA (2 in the data table). One must know these numbers in order to estimate the actual number of *try⁺tyr⁺* cotransformations.

22. (a) Rifampicin eliminates the donor strain which is *rif*ˢ.

(b)   *b a*   *c*   F

(c) To determine the location of the *rif* gene one could use a donor strain which was *rif*ʳ but sensitive to another antibiotic (ampicillin for example). The interrupted mating experiment is conducted as usual on an ampicillin-containing medium but the recombinants must be replated on a rifampicin medium to determine which ones are sensitive.

Chapter 16: Recombinant DNA Technology

Concept Areas	Corresponding Problems
Overview	1, 5, 9, 10
Making DNA Clones	2, 11, 12
Cloning DNA in E. coli	3, 6
Constructing DNA Libraries	4, 7, 8, 14, 15
Identifying Specific Cloned Sequences	12
Methods of Analysis of Cloned Sequences	12, 13, 17, 19, 20
Application	16, 18

Vocabulary: Organization and Listing of Terms and Concepts

Structures and Substances

Recombinant DNA, Clone

Restriction endonucleases

  palindrome

  "sticky" ends

  Eco R1, SmaI

  terminal deoxynucleotidyl transferase

Vector

  cloning vehicle, plasmids

    pUC18

      multiple cloning polylinker site

      lacZ, X-gal

    bacteriophage, λ, M13

  cosmids

    shuttle vector

Bacterial artificial chromosome (BAC)

Yeast artificial chromosome (YAC)

  autonomously replicating sequence (ARS)

Agrobacterium tumifaciens

  Ti plasmid

  opines

Liposome

Cloned DNA fragments

Probe

Poly dA

Poly dT

Replicative form (RF)

Reverse transcriptase

Chapter 16 Recombinant DNA Technology

Oligonucleotide

    primers

      Taq polymerase

        gene specific, random

Restriction map

Dideoxynucleotide

Processes/Methods

  Recombinant DNA technology

    gene splicing

    genetic engineering

      restriction endonucleases

      vector

        plasmids

        selection (antibiotic resistance)

        bacteriophage

          transfection

          DNA sequencing

        cosmids

          cos sequences (lambda)

        shuttle vector

          cloning

        amplification

        antibiotic resistance

    hosts

      E. coli K12 transformation

polymerase chain reaction (PCR)

library construction

    genomic libraries

      $N = \ln(1\text{-}P)/\ln(1\text{-}f\,)$

    chromosome-specific libraries

      subgenomic fraction

flow cytometry

pulse field gel electropohoresis

    cDNA libraries

      reverse transcriptase

      DNA polymerase I

    selection of recombinant clones

      probes

      reverse translation

      chemiluminescence

      colony and plaque hybridization

        nitrocellulose or nylon filter

Analytical methods

  chromosome walking

    open reading frame (ORF)

    restricting mapping

      restriction fragment length polymorphism (RFLP)

    gel electrophoresis

Chapter 16 Recombinant DNA Technology

nucleic acid blotting

   Southern blot

   northern blot

   western blot

DNA sequencing

   applications

   gene mapping

PCR analysis

   denaturation

   annealing of primers

   extension of primers

     heat stable polymerase

     Taq polymerse

yeast

   2 micron plasmid

   yeast artificial chromosome (YAC)

plants

   Agrobacterium tumifaciens

   tumor-inducing plasmid (Ti)

   T-DNA

     callus

     transgenic

mammals

Concepts

Cloning

   selection strategies

Polymerase chain reaction

Probes

Chromosome walking

Restriction mapping

Gene mapping

Gene engineering

   research

   clinical applications

Transgenic organism

Chapter 16 Recombinant DNA Technology

Solutions to Problems and Discussion Questions

1. Recombinant DNA technology, also called genetic engineering or gene splicing, involves the creation of associations of DNA that are not typically found in nature. Particular enzymes, called *restriction endonucleases*, cut DNA at specific sites and often yield "sticky" ends for additional interaction with DNA molecules cut with the same class of enzyme.

Isolated from bacteria, restriction enzymes fall into two classes, type I and type II. *Terminal transferase* is an enzyme which extends single-stranded ends with the addition of "tails" which provide "sticky" ends. A vector may be a plasmid, bacteriophage, or cosmid which receives, through ligation, a piece, or pieces of foreign DNA. The recombinant vector can transform (or transfect) a host cell (bacterium, yeast cell, *etc.*) and be amplified in number. *Calcium chloride* is often used to increase the permeability of host cells to transformation.

2. *Reverse transcriptase* is often used to promote the formation of cDNA (complementary DNA) from a mRNA molecule. Eukaryotic mRNAs typically have a 3' polyA tail as indicated in the diagram below. The poly dT segment provides a double-stranded section which serves to prime the production of the complementary strand.

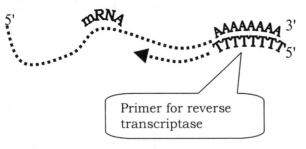

Primer for reverse transcriptase

3. (a) Because the *Drosophila* DNA has been cloned into the *Eco R1* site in the ampicillin resistance gene of the plasmid, the gene will be mutated and any bacterium with the plasmid will be ampicillin sensitive. The tetracycline resistance gene remains active however. Bacteria which have been transformed with the recombinant plasmid will be resistant to tetracycline and therefore tetracycline should be added to the medium. **(b)** Colonies which grow on a tetracycline medium should be tested for growth on an ampicillin medium either by replica plating or some similar controlled transfer method. Those bacteria which do not grow on the ampicillin medium probably contain the *Drosophila* DNA insert.

(c) Resistance to both antibiotics by a transformed bacterium could be explained in several ways. First, if cleavage with the *Eco R1* was incomplete, then no change in biological properties of the uncut plasmids would be expected. Also, it is possible that the cut ends of the plasmid were ligated together in the original form with no insert.

4. Apply the formula:

$$N = \ln(1-P)/\ln(1-f)$$

$$= \ln(1- 0.99)/\ln[1- (5 \times 10^3/1.5 \times 10^8)]$$

$$= \ln(.01)/\ln(.9999667)$$

$$= -4.605/-0.0000333$$

$$= 1.38 \times 10^5$$

5. The question of protein/DNA recognition and interaction is a difficult one to answer. Much research has been done to attempt to understand the nature of the specificity of such interactions. In general it is believed that the protein interacts with the major groove of the DNA helix. This information comes from the structure of the few proteins which have been sufficiently well studied to suggest that the DNA major groove and "fingers" or extensions of the protein form the basis of interaction.

182

6. Given that there is only one site for the action of *Hind*III, then the following will occur. Cuts will be made such that a four base single-stranded set of sticky ends will be produced. For the antibiotic resistance to be present, the ligation will reform the plasmid into its original form. However, two of the plasmids can join to form a dimer as indicated in the diagram below.

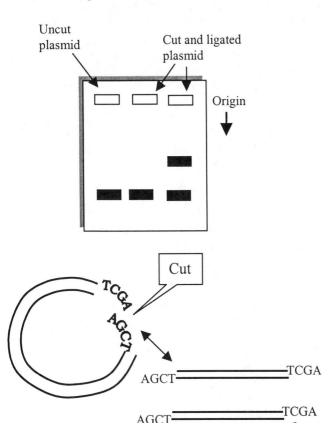

Ligation of two plasmids to make a dimer. Dimer not drawn to scale.

7. Because of complementary base pairing, the 3' end of the DNA strand often loops back onto itself, thereby providing a primer for DNA polymerase I.

8. All other factors being equal (appropriate cloning sites and selectable markers), it is important to consider the size of the foreign DNA which can be cloned into the vector. Generally for large genomes, it is best to use a vector which will accept relatively large fragments.

9. This segment contains the palindromic sequence of GGATCC which is recognized by the restriction enzyme *Bam*HI. The double-stranded sequence is the following

CCTAGG
GGATCC

10. Assuming a random distribution of all four bases, the four-base sequence would occur (on average) every 256 base pairs (4^4), the six-base sequence every 4096 base pairs (4^6), and the eight-base sequence every 65,536 base pairs (4^8). One might use an eight-base restriction enzyme to produce a relatively few large fragments. If one wanted to construct a eukaryotic genomic library, such large fragments would have to be cloned into special vectors, such as yeast artificial chromosomes.

11. A typical procedure is outlined in the K/C text. A filter is used to bind the DNA from the colonies and a labeled probe is used to detect, through hybridization, the DNA of interest. Cells with the desired clone are then picked from the original plate and the plasmid is isolated from the cells.

12. The problem can be best solved by drawing out the strands, then placing the restriction sites in the appropriate positions as follows:

enzyme I   __350_|___950_____

enzyme II 200|_____1100_____

To determine the orientation of the restriction sites to each other, examine the results of the double digested DNA and note that there is a 150bp fragment meaning that enzyme II cuts within the 350bp fragment of enzyme I. Therefore the final map is as follows:

```
        II   I
   200 _|_|_____950_____
          150
```

13. There are several reasons, some stemming from the original use of bacteria as the "workhorse" of molecular biologist as well as the ease with which bacteria and yeast can be manipulated. In addition, there is intense interest in understanding the biology of mammalian cells for obvious reasons and such cells have been manipulated in culture for many years. Perhaps one of the most important reasons is the fact that higher plants lack a suitable variety of vectors which are common to the other cell types mentioned above.

14. There may be several factors contributing to the lack of representation of the 5' end of the mRNA. One has to deal with the possibility that the reverse transcriptase may not completely synthesize the DNA from the RNA template. The other reason may be that the 3' end of the copied DNA tends to fold back on itself thus providing a primer for the DNA polymerase. Additional preparation of the cDNA requires some digestion at the folded region. Since this folded region corresponds to the 5' end of the mRNA, some of the message is often lost.

15. Reverse translation is the process of building a DNA strand (probe) from knowledge of the amino acid sequence of a protein. Using the genetic code a DNA strand can be made which can be prepared for cloning into an appropriate vector or amplified by PCR. A variety of labeling techniques can then be used to identify complementary base sequences contained in the genomic library. One must know at least a portion of the amino acid sequence of the protein in order for the procedure to be applied.

Some problems can occur through degeneracy in the genetic code (not allowing construction of an appropriate DNA), pseudogenes in the library (hybridizations with inappropriate fragments in the library), and variability of DNA sequences in the library due to introns (causing poor or background hybridization).

To overcome some of these problems, one can construct a variety of relatively small probes of different types which take into account the degeneracy in the code. By varying the conditions of hybridization (salt and temperature) one can reduce "background" hybridization.

16. Option (b) fits the expectation because both bands in the offspring are found in both parents. Option (d) would be a possibility, however since the primers were stated as being highly polymorphic in humans, it is likely that they are also highly polymorphic in other primates. Thus, the likelihood of such a match is expected to be low in the general population.

17. With repeated sequences in the genome, chromosome walking is complicated because the clone hybridizes to multiple regions. One can "chromosome jump" over repeated sequences. One can block repeats by addition of repetitive DNA.

18. An appropriate sequence would be the following: 3, 1, 7, 8, 9, 6. Note that (2) may be used to make restriction maps for determining the direction of the walk and the relationships of overlapping clones.

19. (a) Starting with zero at the top, the various patterns tell us that there is an E site at 1000bp, an A site at 500bp, and a B site at 2500bp.

(b) Notice that the probe hybridizes consistently to the 2000bp fragment between the A and B restriction sites.

20. (a) The overall size of the fragment is 12kb. From the A + N digest, sites A and N must be 1kb apart. N must be 2kb from an E site. Pattern #5 is the likely choice. Notice that digest A + N breaks up the 6kb E fragment.

(b) By drawing lines though sections that hybridize to the probe, one can see that the only place of consistent overlap to the probe is the 1kb fragment between A and N.

Chapter 17: Genomics, Bioinformatics, and Proteomics

Concept Areas	Corresponding Problems
Genomic Organization	1, 2, 3, 5, 6, 7
Multigene Families	4

Vocabulary: Organization and Listing of Terms and Concepts

Structures and Substances

Genome

    Pseudomonas aeruginosa

Bacterial chromosomes

    DNA

    double-stranded

    circular, linear

    plasmids

    polycistronic

    high density

        approximately one gene/kb

    operons

Viruses

    overlapping genes

Archaea (archaebacteria)

    rDNA sequence comparisons

extremophiles

    circular, double stranded DNA

histones

introns (tRNA genes)

Eukaryote

    variable gene density

    introns

    repetitive sequences

    Caenorhabditis elegans

        6 chromosomes, 20,000 genes

        repetitive DNA

        introns

        heterochromatin

    higher plants

        Arabidopsis thaliana

Chapter 17 Genomics, Bioinformatics, and Proteomics

humans

    3 billion nucleotides

    protein-coding about 5%

    transposable elements (LINE, Alu)

    gene desert

    30,000-40,000 genes

multigene families

    alpha-globin (2)

        zeta

pseudogene

    beta-globin

        paralogous

    intergenic regions

    epsilon

    gamma (Gγ, Aγ)

    delta

    beta

Histone genes

    tandem repeats

    lack introns

    polarity of transcription

    sequence conservation

Inteins

Bacterial proteome

Nuclear pore complex

Processes/Methods

Genomics

Proteomics

Human Genome Project (HGP)

Human Genome Organization (HUGO)

    clone-by-clone

    shotgun method

ELSI (ethical, legal, and social aspects)

TIGR (The Institute for Genome Research)

    annotation

    open reading frame (ORF)

    ATG

    TAA, TAG, TGA

    3'AATAAA poly A signal

    CpG

Genome contraction

Genome evolution

Genome duplication

Gene duplication

    unequal crossing over

    sister chromatid exchange

    unequal sister chromatid exchange

    replication errors

Imprecise joining

Break-nibble-add

Chapter 17 Genomics, Bioinformatics, and Proteomics

Sequence conservation

Proteomics

    peptide mass fingerprinting

Concepts

Bioinformatics

Genomics

Proteomics

Genome organization comparisons

Genome evolution

Minimum genome size (250-350)

Relationships:

    Archaea

    Eubacteria

    Eukaryotes

    multigene families

    globin genes

    immunoglobin genes

    histone genes

Sequence conservation

Solutions to Problems and Discussion Questions

1. General similarities and differences:

Yeast	Bacteria
DNA double-stranded chromosomes 12 MB 6548 genes	DNA double-stranded circular (*E. coli*) naked nucleic acid 4.6 MB 4397 genes

2. While greater DNA content per cell is associated with eukaryotes, one can not universally equate genomic size with an increase in organismic complexity. There are numerous examples where DNA content per cell varies considerably among closely related species. Because of the diverse cell types of multicellular eukaryotes, a variety of gene products is required, which may be related to the increase in DNA content per cell.

In addition, the advantage of diploidy automatically increases DNA content per cell. However, seeing the question in another way, it is likely that a much higher *percentage* of the genome of a prokaryote is actually involved in phenotype production than in a eukaryote.

Eukaryotes have evolved the capacity to obtain and maintain what appears to be large amounts of "extra" perhaps "junk" DNA. This concept will be examined in subsequent chapters of the text. Prokaryotes on the other hand, with their relatively short life cycle, are extremely efficient in their accumulation and use of their genome. Given the larger amount of DNA per cell in eukaryotes and the requirement that the DNA be partitioned in an orderly fashion to daughter cells during cell division, certain mechanisms and structures (mitosis, nucleosomes, centromeres, *etc.*) have evolved for *packaging* the DNA. In addition, the genome is divided into separate entities (chromosomes) to perhaps facilitate the partitioning process in mitosis and meiosis.

3. Bacterial genes are densely packed in the chromosome. The protein-coding genes are mostly organized in polycistronic transcription units without introns. Eukaryotic genes are less densely packed in chromosomes and protein-coding genes are mostly organized as single transcription units with introns.

4. While the β-globin gene family is a relatively large (60kb) sequence and restriction analyses show that it is composed of six genes, one is a pseudogene and therefore does not produce a product. The five functional genes each contain two similarly-sized introns which when included with non-coding flanking regions (5' and 3'), and spacer DNA between genes, accounts for the 95% mentioned in the question.

5. Depending on the reading frame and the stand being read, a relatively large number of open reading frames can exist within a stretch of double-stranded DNA. Because genes tend to be longer than 50 codons, a lower size limit is place on any given sequence being examined for location of an actual gene. The smaller the size limit, the larger the number of ORFs that will be identified as possible genes.

6. Many plants have genomes larger than *Arabidopsis* but have about the same number of genes. Much of the extra DNA in large-genome plants is in the form of repetitive DNA, often transposons. Because plants closely related to *Arabidopsis* often have large genomes, it is thought that the small genome of *Arabidopsis* arose through genomic contraction.

7. In parasitic cells, nutrients are often obtained from the host rather than from *de novo* synthesis. Thus, theoretically, parasites may be able to streamline their genomes if they could effectively use a relatively high proportion of the host's metabolic reactants and products. Viruses, for example, produce progeny with very few genes, however, they complete their replicative activities *within* a cell. Even if parasitic, a cell (not a virus), must be able to interact effectively with its environment, replicate its genome, reproduce, and engage in a variety of metabolic activities. It is likely that a couple of hundred gene products would be required for these activities.

Chapter 18: Applications and Ethics of Genetic Technology

Concept Areas	Corresponding Problems
Mapping Human Genetic Disorders	3, 6, 7
Diagnosing and Screening Genetic Disorders	8, 9, 11, 14, 15
Gene Therapy	1, 4, 10, 13
Genome Analysis	5
Biotechnology	2, 12

Vocabulary: Organization and Listing of Terms and Concepts

Structures and Substances

Recombinant DNA molecules

RFLP markers

  DNA blot (Southern)

  neurofibromastosis (NF1)

  neurofibromin

  exclusion map

  Marfan syndrome

    dystrophin

    fibrillin

Allele-specific oligonucleotides

  cystic fibrosis

  transmembrane conductance regulator (CFTR)

DNA chip

  p53, BRCA1

  retroviral vector

  Moloney virus

Adenosine deaminase

Erythropoetin

Rapamycin

Adeno-associated virus

YAC, BAC

T cells

Alpha-1-antitrypsin

Insulin, proinsulin

EPSP synthase

Fusion proteins (polypeptide)

Transgenic organisms

  biofactories, vaccine

Processes/Methods

Linkage and mapping

  positional cloning

  FISH (fluorescent *in situ* hybridization)

Chapter 18 Applications and Ethics of Genetic Technology

Logarithm of the odds (lod)

    centiMorgan (cM)

        $1\text{-}3 \times 10^6$ bp

Cloning of genes

    positional cloning

    candidate genes

        Marfan syndrome

            fibrillin

Diagnostics

    amniocentesis

    chorionic villus sampling (CVS)

        β-globin

        thalassemia

        sickle cell anemia

            Southern blot

            allele-specific oligonucleotide

            polymerase chain reaction

Gene therapy

    severe combined immunodeficiency (SCID)

        adenosine deaminase (ADA)

        hypercholesterolemia

    somatic gene therapy

    germline gene therapy

    enhancement gene therapy

DNA fingerprinting

    RFLP

    minisatellites

    variable number tandem repeats (VNTR)

    applications

        forensic

Biotechnology

    Celera, PE Biosystems

        proteome

        transcriptome

    commercial applications

        insulin

        preproinsulin

    pharmaceutical products

        milk of livestock

        alpha-1-antitrypsin

Transgenesis

    crops

        herbicide resistance

            Ti vector

            glyphosate resistance

    vaccines

        inactivated

        attenuated

        subunit vaccine

        hepatitis B

Chapter 18 Applications and Ethics of Genetic Technology

Concepts

Gene mapping

    positional cloning

Cloning of genes

Diagnostics

    probing

    ethics

Therapy

    somatic gene therapy

    germline gene therapy

    enhancement gene therapy

    associated problems

    guidelines

Ethical issues

Vector strategies

Human Genome Project (HGP)

ELSI (Ethical, Legal, and Social Implications)

TIGR (The Institute for Genomic Research)

Human Genome Organization (HGO)

Transgenic systems

Vaccine administration

Chapter 18 Applications and Ethics of Genetic Technology

Solutions to Problems and Discussion Questions

1. The nature of the digestion process is the breakdown of foodstuffs for eventual absorption by the small intestine. Antigens are usually quite large molecules, and in the process of digestion, they are sometimes broken down into smaller molecules, thus becoming ineffective in stimulating the immune system.   Some individuals are allergic to the food they eat, testifying to the fact that all antigens are not completely degraded or modified by digestion. In some cases ingested antigens do indeed stimulate the immune system (oral polio vaccine) and provide a route for immunization. Localized (intestinal) immunity can sometimes be stimulated by oral introduction of antigens and in some cases this can offer immunity to ingested pathogens.

2. Glyphosate (a herbicide) inhibits EPSP, a chloroplast enzyme involved in the synthesis of the amino acids phenylalanine, tyrosine, and tryptophan. To generate glyphosate resistance in crop plants a fusion gene was created which introduced a viral promoter to control the EPSP synthetase gene. The fusion product was placed into the Ti vector and transferred to *A. tumifaciens* which was used to infect crop cells. Calluses were selected on the basis of their resistance to glyphosate. Resistant calluses were later developed into transgenic plants.   There is a remote possibility that such an "accident" can occur. However, in retracing the steps to generate the resistant plant in the first place, it seems more likely that the trait will not "escape" from the plant; rather that the engineered *A. tumifaciens* may escape, infect and transfer glyphosate resistance to pest species.

3. Enhancement gene therapy opens the door to a variety of ethical issues. What limits can/should be imposed on individuals or institutions seeking to improve human qualities? What qualities should be open for enhancement? Gene therapy is not without medical and ethical risks.

4. (a,b) One of the main problems with gene therapy is delivery of the desired virus to the target tissue in an effective manner. Several of the problems involving the use of retroviral vectors are the following. (1) Integration into the host must be cell specific so as not to damage non-target cells. (2) Retroviral integration into host cell genomes only occurs if the host cell is replicating. (3) Insertion of the viral genome might influence non-target but essential genes. (4) Retroviral genomes have a low cloning capacity and can not carry large inserted sequences as are many human genes. (5) There is a possibility that recombination with host viruses will produce an infectious virus which may do harm.

(c) The question posed here plays on the practical versus the ethical. It would certainly be more efficient (although perhaps more difficult technically) to engineer germ tissue, for once it is done in a family, the disease would be eliminated. However, there are considerable ethical problems associated with germ plasm therapy. It recalls previous attempts of the eugenics movements of past decades which involved the use of selective breeding to purify the human stock. Some present-day biologists have said publically that germ line gene therapy will *not* be conducted.

5. *Drosophila* is a unique experiemental organism in that there is a vast knowledge of its genetics, it is easily cultured and genetically manipulated, and it contains unique chromosomes, polytene chromosomes, which allow visual landmarks. Coupled with probe-labeling (sequence tagged sites), the visible landmarks (chromomeres) and ease of manipulation, one can actually see where important genes are located in chromosomes. *Drosophila* also contains P elements which allow sequence markers to be inserted into the genome. Microdissection of chromosomes is also useful in developing specific clones for sequencing. In addition, techniques have been developed (*in situ* hybridization) to allow scientists to actually determine the distributions of gene activities in all tissues of the organism.

193

6. Positional cloning is a technique whereby the linkage group of a genetic disorder is determined by association of certain RFLP (as markers). A more precise location is obtained by association (linkage studies in kindreds) with additional RFLP markers. Once the general region of the gene is located, sequencing is employed to identify the genes within the general region. Comparison of the sequences in individuals with and without the affliction allows the identification of the gene responsible for the disease. Difficulties in positional cloning relate to the availability of sufficient RFLP markers in the region of the gene and sufficient kindreds to do the actual genetic association of the gene in question to the RFLP markers. Mutations which would speed the process would be those which are most easily seen, perhaps those which change the banding patterns of chromosomes (deletions, duplication, inversions. translocations) or those which cause sufficient base changes that would alter probe hybridization.

7. Positional cloning relies on segregation (Mendelian) and linkage analysis. Given the numerous limitations associated with such analyses in human populations (family and sample size, *etc.*) it is unlikely that this technique will be successfully applied to genetically complex traits in the near future.

8. Even though you have developed a method for screening seven of the mutations described, it is possible that negative results can occur even though the person carries the gene for CF. In other words, the specific probes (or allele-specific oligonucleotides) that have been developed will not necessarily be useful for screening all mutant genes. In addition, the cost-effectiveness of such a screening proposal would need to be considered.

9. In the case of haplo-insufficient mutations, gene therapy holds promise; however in "gain-of - function" mutations in all probability, the mutant gene's activity or product must be compromised. Addition of a normal gene probably will not help.

10. It will hybridize by base complementation to the normal DNA sequence.

11. The answer provided here is based on the condition that individual I-2 is a carrier and the son, II-4, has the disorder. The 3kb fragment occurs in the normal I-1 father and the normal son II-1. The affected son, II-4, has the 4kb fragment. One daughter, II-2, is a carrier while the other daughter, II-3, is not a carrier.

12. One method is to use the amino acid sequence of the protein to produce the gene synthetically. Alternatively, since the introns are spliced out of the hnRNA in the production of mRNA, if mRNA can be obtained, it can be used to make DNA (cDNA) through the use of reverse transcriptase.

13. The two major problems described here are common concerns related to genetic engineering. The first is the localization of the introduced DNA into the target tissue and target location in the genome. Inappropriate targeting may have serious consequences. In addition, it is often difficult to control the output of introduced DNA. Genetic regulation is complicated and subject to a number of factors including upstream and downstream signals as well as various posttranscriptional processing schemes. Artificial control of these factors will prove difficult.

14. The child in question is a carrier of the deletion in the beta-globin gene, just as the parents are carriers. Its genotype is therefore $\beta^A\beta^o$.

15.

(a) Y-linked excluded, X-linked recessive excluded, autosomal recessive possible but unlikely, X-linked dominant possible if heterozygous, autosomal dominant possible.

(b) Chromosome 21 with the B1 marker probably contains the mutation.

(c) The disease gene is segregating with some certainty with the B1 RFLP marker in the family. Since the mother also has the B3 marker, the offspring could be tested. If the child carries the B3 marker, then he/she does not carry the B1 marker which has been segregating with the defective gene. However, this prediction is not completely accurate because a crossover in the mother could put the undesirable gene with the B3 marker.

(d) A crossover between the restriction sites in the father giving a B1 chromosome, or a mutation eliminating either the B2 or B3 restriction site.

Chapter 19: Regulation of Gene Expression in Prokaryotes

Concept Areas	Corresponding Problems
Overview	1, 10, 11, 12
Lactose Metabolism in E. coli: Inducible	1, 3, 4, 5, 6, 7, 8, 9
Positive and Negative Control	2, 12, 13
Tryptophan Operon	1, 3, 14 15

Vocabulary: Organization and Listing of Terms and Concepts

Structures and Substances

Lactose

  structural genes

  lac operon

    cis-acting

    trans-acting

    lac Z

      β-galactosidase

    lac Y

      permease

    lac A

      transacetylase

  polycistronic mRNA

  gratuitous inducers

    isopropylthiogalactoside (IPTG)

  constitutive mutants

    lac I

    lac O^c

merozygote

regulatory units

  repressor gene

    repressor molecule

    repression loop

    diffusible cellular product (F19.2)

    operator region

    no diffusible product

    adjacent control (F19.2)

  lac I^s

  lac I^q

catabolite activating protein (CAP)

  promoter region

  glucose

  CAP binding site

  cyclic adenosine monophosphate (cAMP)

  adenyl cyclase

Chapter 19 Regulation of Gene Expression in Prokaryotes

Arabinose

   *ara*B, A, D, I, O$_2$

   *ara*C regulatory protein

Tryptophan

   tryptophan synthetase

   trp R$^-$, *trp* R$^+$

     co-repressor

   structural genes

     trp E, D, C, B, A

   trp P-*trp* O region

   leader sequence

     attenuator

     tRNA^{trp}

     termination loop

     anti-termination hairpin

     Bacillus subtilis

     TRAP (trpRNA-binding attenuating protein)
     AT (anti-TRAP)

Processes/Methods

Genetic regulation

  adaptive

   inducible

    inducer

    lactose

  equilibrium dialysis

   catabolite repression

   cooperative binding

  repressible

   tryptophan

    attenuation

     ribosome "stall"

    negative, positive control

     catabolite repression

    arabinose regulation

  constitutive

  allosteric

  superrepression

Concepts

Genetic regulation

  efficiency

Cis-acting, *Trans*-acting

Positive control (F19.1)

  catabolite repression

Negative control (F19.1)

  lactose operon

  tryptophan operon

   repression

   attenuation

   "stalling"

F19.1 Illustration of general processes of *negative* and *positive* control. If *negative* control is operating, the regulatory protein inhibits transcription. With *positive* control, transcription is stimulated.

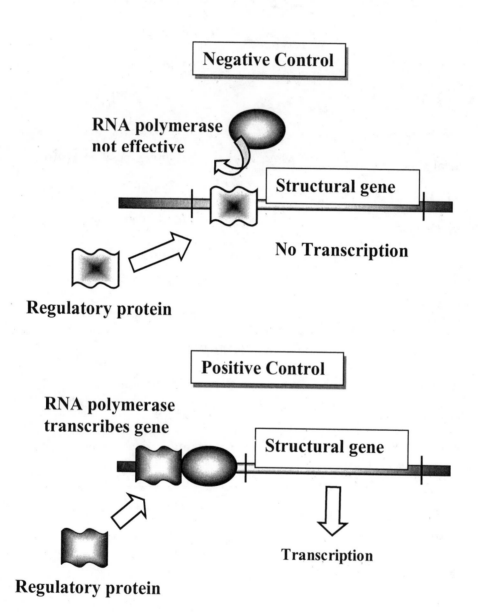

F19.2 Illustration of the nature of the product of the *I* gene. It can act "at a distance" because it is a protein which can diffuse through the cytoplasm and thus act in *trans*. There is no protein product of the operator gene, therefore it can only act in *cis*.

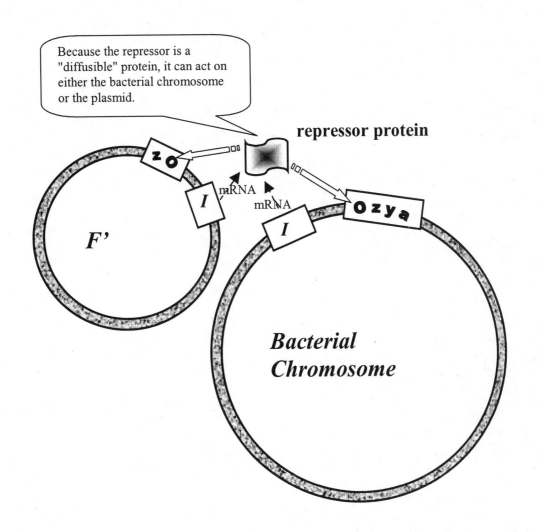

Solutions to Problems and Discussion Questions

1. The answer to this question is a key to enhancing a student's understanding of the Jacob-Monod model as related to lactose and tryptophan metabolism. The enzymes of the lactose operon are needed to break down and use lactose as an energy source. If lactose is the sole carbon source, the enzymes are synthesized to use that carbon source. With no lactose present, there is no "need" for the enzymes.

The tryptophan operon contains structural genes for the *synthesis* of tryptophan. If there is little or no tryptophan in the medium, the tryptophan operon is "turned on" to manufacture tryptophan. If tryptophan is abundant in the medium, then there is no "need" for the operon to be manufacturing "tryptophan synthetases."

2. Refer to F19.1 to see that under *negative* control, the regulatory molecule interferes with transcription while in *positive* control, the regulatory molecule stimulates transcription. Negative control is seen in the *lactose* and *tryptophan* systems as well as a portion of the *arabinose* regulation. Catabolite repression and a portion of the *arabinose* regulatory systems are examples of positive control.

3. In an *inducible system*, the repressor which normally interacts with the operator to inhibit transcription, is inactivated by an *inducer*, thus permitting transcription. In a *repressible system*, a normally inactive repressor is *activated* by a *co-repressor*, thus enabling it (the activated repressor) to bind to the operator to inhibit transcription. Because the interaction of the protein (repressor) has a negative influence on transcription, the systems described here are forms of *negative control* (see F19.1).

4. (a) Due to the deletion of a base early in the *lac* Z gene there will be "frameshift" of all the reading frames downstream from the deletion. It is likely that either premature chain termination of translation will occur (from the introduction of a nonsense triplet in a reading frame) or the normal chain termination will be ignored. Regardless, a mutant condition for the Z gene will be likely. If such a cell is placed on a lactose medium, it will be incapable of growth because β-galactosidase is not available. **(b)** If the deletion occurs early in the A gene, one might expect impaired function of the A gene product, but it will not influence the use of lactose as a carbon source.

5. Refer to the K/C text and to F19.1,2 to get a good understanding of the lactose system before starting.

$I^+ O^+ Z^+$ = **Inducible** because a repressor protein can interact with the operator to turn off transcription.

$I^- O^+ Z^+$ = **Constitutive** because the repressor gene is mutant, therefore no repressor protein is available.

$I^+ O^c Z^+$ = **Constitutive** because even though a repressor protein is made, it can not bind with the mutant operator.

$I^- O^+ Z^+ / F' I^+$ = **Inducible** because even though there is one mutant repressor gene, the other I^+ gene, on the F factor, produces a normal repressor protein which is diffusible and capable of interacting with the operon to repress transcription. (See F17.2 in this book)

$I^+ O^c Z^+ / F' O^+$ = **Constitutive** because there is a constitutive operator (O^c) next to a normal Z gene. Remembering that this operator functions incis and is not influenced by the repressor protein, constitutive synthesis of β-galactosidase will occur.

$I^s\ O^+\ Z^+$ = **Repressed** because the product of the is gene is *insensitive* to the inducer lactose and thus can not be inactivated. The repressor will continually interact with the operator and shut off transcription regardless of the presence or absence of lactose.

$I^s\ O^+\ Z^+\ /F'\ I^+$ = **Repressed** because, as in the previous case, the product of the I^s gene is insensitive to the inducer lactose and thus can not be inactivated. The repressor will continually interact with the operator and shut off transcription regardless of the presence or absence of lactose. The fact that there is a normal I^+ gene is of no consequence because once a repressor from I^s binds to an operator, the presence of normal repressor molecules will make no difference.

6. Refer to the K/C text and to F19.2 to get a good understanding of the lactose system before starting.

$I^+\ O^+\ Z^+$ = Because of the function of the active repressor from the I^+ gene, and no lactose to influence its function, there will be **No Enzyme Made**.

$I^+\ O^c\ Z^+$ = There will be a **Functional Enzyme Made** because of the constitutive operator is in *cis* with a Z gene. The lactose in the medium will have no influence because of the constitutive operator. The repressor can not bind to the mutant operator.

$I^-\ O^+\ Z^-$ = There will be a **Nonfunctional Enzyme Made** because with I^- the system is constitutive but the Z gene is mutant. The absence of lactose in the medium will have no influence because of the non-functional repressor. The mutant repressor can not bind to the operator.

$I^-\ O^+\ Z^-$ = There will be a **Nonfunctional Enzyme Made** because with I^- the system is constitutive but the Z gene is mutant. The lactose in the medium will have no influence because of the non-functional repressor. The mutant repressor can not bind to the operator.

$I^-\ O^+\ Z^+\ /F'\ I^+$ = There will be **No Enzyme Made** because in the absence of lactose, the repressor product of the I^+ gene will bind to the operator and inhibit transcription.

$I^+\ O^c\ Z^+\ /F'\ O^+$ = Because there is a constitutive operator in *cis* with a normal Z gene, there will be **Functional Enzyme Made**. The lactose in the medium will have no influence because of the mutant operator.

$I^+\ O^+\ Z^-\ /F'\ I^+\ O^+\ Z^+$ = Because there is lactose in the medium, the repressor protein will not bind to the operator and transcription will occur. The presence of a normal Z gene allows a **Functional and Non-functional Enzyme to be Made**. The repressor protein is diffusable, working in *trans*.

$I^-\ O^+\ Z^-\ /F'\ I^+\ O^+\ Z^+$ = Because there is no lactose in the medium, the repressor protein (from I^+) will repress the operators and there will be **No Enzyme Made**.

$I^s\ O^+\ Z^+\ /\ F'\ O^+$ = With the product of I^s there is binding of the repressor to the operator and therefore **No Enzyme Made**. The lack of lactose in the medium is of no consequence because the mutant repressor is insensitive to lactose.

$I^+\ O^c\ Z^+\ /F'\ O^+\ Z^+$ = The arrangement of the constitutive operator (O^c) with the Z gene will cause a **Functional Enzyme to be Made**.

7. A single *E. coli* cell contains very few molecules of the *lac* repressor. However, the *lac* I^q mutation causes a 10X increase in repressor protein production, thus facilitating its isolation. With the use of dialysis against a radioactive gratuitous inducer (IPTG), Gilbert and Muller-Hill were able to identify the repressor protein in certain extracts of *lac* I^q cells. The material which bound the labeled IPTG was purified and shown to be heat labile and have other characteristics of protein. Extracts of *lac* I^- cells did not bind the labeled IPTG.

The IPTG-binding protein was labeled with sulfur-containing amino acids and mixed with DNA from λ phage which contained the *lac O*⁺ section of DNA. By glycerol gradient centrifugation it was shown that the labeled repressor protein binds only to DNA which contained the *lac O*⁺ region, thus indicating a specific binding to DNA.

8. In order to understand this question, it is necessary that you understand the negative regulation of the *lactose* operon by the *lac* repressor as well as the positive control exerted by the CAP protein. Remember, if lactose is present, it inactivates the *lac* repressor. If glucose is present, it inhibits adenyl cyclase thereby reducing, through a lowering of cAMP levels, the positive action of CAP on the *lac* operon.

(a) With no lactose and no glucose, the operon is off because the *lac* repressor is bound to the operator and although CAP is bound to its binding site, it will not override the action of the repressor.

(b) With lactose added to the medium, the *lac* repressor is inactivated and the operon is transcribing the structural genes. With no glucose, the CAP is bound to its binding site, thus enhancing transcription.

(c) With no lactose present in the medium, the *lac* repressor is bound to the operator region, and since glucose inhibits adenyl cyclase, the CAP protein will not interact with its binding site. The operon is therefore "off."

(d) With lactose present, the *lac* repressor is inactivated, however since glucose is also present, CAP will not interact with its binding site. Under this condition transcription is severely diminished and the operon can be considered to be "off."

9. (a) Because activated CAP is a component of the cooperative binding of RNA polymerase to the *lac* promoter, absence of a functional *crp* would compromise the positive control exhibited by CAP.

(b) Without a CAP binding site there would be a reduction in the inducibility of the *lac* operon.

10. First notice that in the first row of data, the presence of tm in the medium causes the production of active enzyme from the wild type arrangement of genes. From this one would conclude that the system is *inducible*. To determine which gene is the structural gene, look for the *IE* function and see that it is related to *C*. Therefore *C* codes for the **structural gene**. Because when *B* is mutant, no enzyme is produced, *B* must be the **promoter**.

Notice that when genes *A* and *D* are mutant, constitutive synthesis occurs, therefore one must be the operator and the other gene codes for the repressor protein. To distinguish these functions, one must remember that the repressor operates as a diffusible substance and can be on the host chromosome or the F factor (functioning in *trans*). However, the operator can only operate in *cis*. In addition, in *cis*, the constitutive operator is dominant to its wild type allele, while the mutant repressor is recessive to its wild type allele.

Notice that the mutant *A* gene is dominant to its wild type allele, whereas the mutant *d* allele is recessive (behaving as wild type in the first row). Therefore, the *A* locus is the **operator** and the *D* locus is the **repressor** gene.

11. Because the deletion of the regulatory gene causes a loss of synthesis of the enzymes, the regulatory gene product can be viewed as one exerting *positive control*. When tis is present, no enzymes are made, therefore, tis must inactivate the positive regulatory protein. When tis is absent, the regulatory protein is free to exert its positive influence on transcription. Mutations in the operator negate the positive action of the regulator. On the next page (F19.3) is a model which illustrates these points.

F19.1 Model of regulatory system described in problem #11. This is an example of *positive* control.

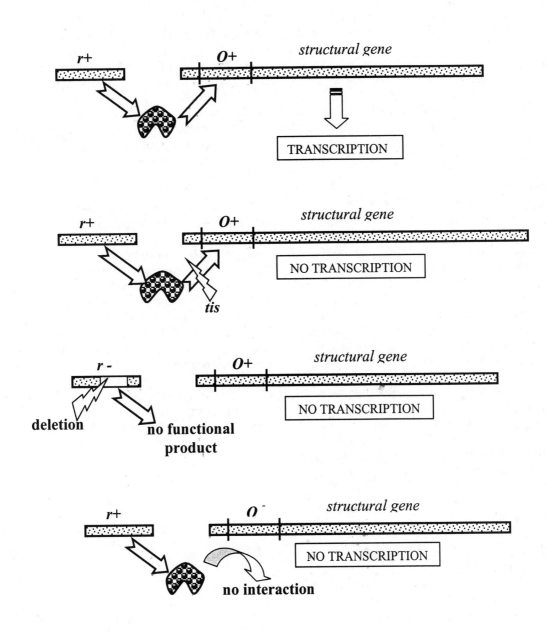

12. The first two sentences in the problem indicate an inducible system where oil stimulates the production of a protein(?) which turns on (positive control) genes to metabolize oil. The different results in strains #2 and #4 suggest a *cis*-acting system. Because the operon by itself (when mutant as in strain #3) gives constitutive synthesis of the structural genes, *cis*-acting system is also supported. The *cis*-acting element is most likely part of the operon.

13. (a) Call one constitutive mutation *lexA*- (mutation in the repressor gene product) and the other O^{uvrA-} (mutation in the operator).

(b) One can make partial diploid strains using F'. O^{uvrA-} will be dominant to O^{uvrA+} and *lexA*- will be recessive to *lexA*+. O^{uvrA-} will act in *cis*.

(c) If one could develop an assay for the other gene products under SOS control, with a *lexA*- strain the other gene products should be present at induced levels.

14. You will need to identify the complementary regions. This is one of the rare cases in this handbook where the entire answer will not be directly given. You will find four regions which "fit." To get started, find the CACUUCC sequence. It pairs, with one mismatch with a second region. Hint: The third region is composed of seven bases and starts with an AG.

15. (a) With tryptophan being abundant, the assumption is that charged $tRNA^{trp}$ is also present. TRAP should be saturated with tryptophan and be actively bound to the 5' end of the nascent mRNA. Therefore, the structural genes should not be expressed.

(b) If tryptophan is scarce, even though $tRNA^{trp}$ is present it should not be charged. TRAP is present but it is not saturated with tryptophan. The structural genes should be expressed.

(c) The answer to this situation needs some qualification. If tryptophan is abundant, but $tRNA^{trp}$ is scarce, the assumption is that charged $tRNA^{trp}$ is being described. This could be due to a nonfunctional tryptophanyl-tRNA synthetase. In that case, the uncharged $tRNA^{trp}$ can induce AT, which then binds to tryptophan-saturated TRAP. This prevents TRAP from binding to the leader RNA sequence, thus allowing expression of the tryptophan operon.

(d) With no TRAP, there can be no termination of transcription. Therefore under this condition, even with abundant tryptophan, there is expression of the operon.

Chapter 20: Regulation of Gene Expression in Eukaryotes

Concept Areas	Corresponding Problems
Overview	1, 2, 3, 7
Regulatory Elements	3, 6
Transcription Factors	4, 5, 9
Promoters, Enhancers	4
Posttranscriptional Regulation	8

Vocabulary: Organization and Listing of Terms and Concepts

Structures and Substances

Nuclear territories

   interchromosomal domains

      channels

      CD4 locus

      T cells

   gamma satellite

   territory borders

Non-histone proteins

   nucleosome

Chromatin

   N-terminal tail

   remodeling enzyme

   remodeling machine

Promoters

Introns, exons

Histones

Transcription factors

   positive/negative factors

   functional domains

      DNA-binding domains

      trans-activating domains

      motif, GAL4, UAS_G

   TFIID, TFIIB, TFIIF, TFIIIA

   TBP, TAFs (TATA), others

   zinc fingers

   helix-turn-helix (HTH)

      homeobox, homeodomain

   basic leucine zippers (bZIP)

   leucine zipper

RNA polymerase I, II, III

   sigma factor

Promoters

Enhancers

    positive

    upstream activator sequences (UAS)

    antirepressor

    (SWI/SNF)

    HAT   (histone acetyltransferase)

    deacetylase (HDAC)

    insulator element

    Transcription complex

        5'-azacytidine, globin genes

    Enhansons

    Transcripts

        introns, exons

        preprotachykinin mRNA (PPT)

        tachykinin

        sex-determining genes in

          Drosophila (Sxl, tra, dsx)

Processes/Methods

Chromosome painting

Chromosome remodeling

Epigenetic processes

Histone acetylation

Gene regulation

    transcription

        RNA polymerase I, II, III

upstream (5') organization

    promoters

        TATA box (-25 to -30)

        AT rich

        GC flanking

        CAAT box (-70 to -80)

        or CCAAT

        GC box (-110)

    enhancers

        variable position

        variable orientation

        cis-acting

        upstream and downstream

          upstream activator

          sequences (UAS)

        transcription factors

        functional domains

promoter clearance

basal, induced states

    positive, negative

    antirepressors

ATP-hydrolysis dependent remodeling

DNA binding

    trans-activating

        structural motifs

Chapter 20 Regulation of Gene Expression in Eukaryotes

zinc fingers

homeobox

homeodomain (HD)

leucine zipper

activator domains

galactose metabolism

catabolite repression

positive control

DNase hypersensitivity

gene alterations

DNA methylation

tissue specificity

processing (post-transcriptional regulation)

intron removal/exon splicing

alternative splicing

3'-polyadenylation (polyA tail)

transport

mRNA stability

stability sequence

address sequence

instability element

ongogene

translational control

autoregulation

tubulin regulation

RNAse action and degradation

posttranslational modification

Concepts

Comparison to prokaryotes

Genetic regulation

methylation

transcription

processing

alternative processing

transport

stability

translation

posttranslational modification

Stimulation of transcription

positive inducibility

positive control

catabolite repression

relationships to transcription factors

relationships to enhancers

Autoregulation

Sex-determination in *Drosophila*

Chapter 20 Regulation of Gene Expression in Eukaryotes

Solutions to Problems and Discussion Questions

1. There are several reasons for anticipating a variety of different regulatory mechanisms in eukaryotes as compared to prokaryotes. Eukaryotic cells contain greater amounts of DNA and this DNA is associated with various proteins, including histones and nonhistone chromosomal proteins. *Chromatin* as such does not exist in prokaryotes. In addition, whereas there is usually only one chromosome in prokaryotes, eukaryotes have more than one chromosome all enclosed in a membrane (nuclear membrane). This nuclear membrane separates, both temporally and spatially, the processes of transcription and translation thus providing an opportunity for post-transcriptional, pre-translational regulation.

While prokaryotes respond genetically to changes in their external environment, cells of multicellular eukaryotes interact with each other as well as the external environment. The structural and functional diversity of cells of a multicellular eukaryote, coupled with the finding that all cells of an organism contain a complete complement of genes, suggests that in some cells certain genes are active which are not active in other cells.

It is often difficult to study eukaryotic gene regulation because of the complexities mentioned above, especially tissue specificity and the various levels at which regulation can occur (as indicated in question #2 below). Obtaining a homogeneous group of cells from a multicellular organism often requires a significant alteration of the natural environment of the cell. Thus, results from studies on isolated cells must be interpreted with caution. In addition, because of the variety of intracellular components (nuclear and cytoplasmic) it is difficult to isolate, free of contamination, certain molecular species. Even if such isolation is accomplished, it is difficult to interpret the actual behavior of such molecules in an artificial environment.

2. *Chromatin remodeling*: Changes in DNA/chromosome structure can influence overall gene output. DNA methylation also influences transcription efficiency.

While not specifically discussed in this chapter, *gene amplification* refers to cases where an increase in gene products is achieved by an increase in the number of genes producing those products. Such amplification can be achieved intrachromosomally (chorion genes) or extrachromosomally (rRNA genes in some amphibians).

Transcription: There are several factors which are known to influence transcription: *promoters*, TATA, CAAT, and GC boxes, as well as other upstream regulatory sequences; *enhancers*, which are *cis*-acting sequences that act at various locations and orientations; *transcription factors*, with various structural motifs (zinc fingers, homeodomains, and leucine zippers) which bind DNA and influence transcription; *receptor-hormone complexes* which influence transcription.

Processing and transport types of regulation involve the efficiency of hnRNA maturation as related to capping, polyA tail addition, intron removal, mRNA stability.

Translation: After mRNAs are produced from the processing of hnRNA, they have the potential of being translated. The stability of the mRNAs appears to be an additional regulatory control point. Certain factors, such as protein subunits may influence a variety of steps in the translational mechanism. For instance, a protein or protein subunit may activate an RNase which will degrade certain mRNAs or a particular regulatory element may cause a ribosome to stall, thus decreasing the speed of translation and increasing the exposure of a mRNA to the action of RNAses.

3. *Promoters* are conserved DNA sequences which influence transcription from the "upstream" side (5') of mRNA coding genes. They are usually fixed in position and within 100 base pairs of the initiation site for mRNA synthesis. Examples of such promoters are the following: TATA, CAAT, and GC boxes.

Enhancers are *cis*-acting sequences of DNA which stimulate the transcription from most, if not all, promoters. They are somewhat different from promoters in that the position of the enhancer need not be fixed; it may be upstream, downstream, or within the gene being regulated. The orientation may be inverted without significantly influencing its action. Enhancers can work on different genes, that is, they are not gene-specific.

4. Transcription factors are proteins which are *necessary* for the initiation of transcription. However, they are not *sufficient* for the initiation of transcription. To be activated, RNA polymerase II requires four or five transcription factors. Transcription factors contain at least two functional domains: one binds to the DNA sequences of promoters and/or enhancers, the other interacts with RNA polymerase or other transcription factors.

5. True activators are modular proteins with at least two domains. One domain binds to DNA and the other interacts with RNA polymerase or other transcription factors. Antirepressors remodel chromatin and make DNA open for transcription.

6. Both the *lac* and *gal* systems are influenced by catabolite repression, however, the *lac* system is under negative control whereas the *gal* system is under positive control. Both systems are inducible.

7. Your essay should deal with the following issues:

- differences in basic chromosome structure
- chromosome remodeling
- histone acetylation
- differences in gene structure
- "C" value paradox and its implications
- cell structure (nucleus in eukaryotes)
- levels of potential regulation
  . . .transcriptional
  . . .mRNA processing
  . . .transport
  . . .selection for processing and translation
  . . .mRNA stability
  . . .posttranslational processing
- genomic aspects (amplification, *etc.*)
- biological context in terms of multicellular interactions *versus* single cell survival

8. The work of Cleveland and colleagues allowed selective changes to be made in the *met-arg-glu-lys* sequence. Only the engineered mRNA sequences which caused an amino acid substitution negated the autoregulation, indicating that it is the sequence of the amino acids, not the mRNA which is critical in the process of autoregulation. Notice that code degeneracy allows for changes in mRNA sequence without changes in the amino acid sequence. Therefore, the model which depicts binding of factors to the nascent polypeptide chain is supported. There are a variety of experiments which could be used to substantiate such a model. One might stabilize the proposed MREI-protein complex with "crosslinkers," treat with RNAse to digest mRNA and to break up polysomes, then isolate individual ribosomes. One may use some specific antibody or other method to determine whether tubulin subunits contaminate the ribosome population.

9. (a) There is no place for the TFIID to bind.

(b) There is more transcription in the nuclear extracts. Perhaps other factors not present in the purified system are present in the nuclear extracts.

(c) There is a region probably in the -81 to -50 area which responds to a component in the nuclear extract to bring about high efficiency transcription.

Chapter 21: Developmental Genetics

Concept Areas	Corresponding Problems
Developmental Concepts	1, 9
Variable Gene Activity Theory	4
Differential Transcription in Development	3, 4, 5
Genetics of Embryonic Development	2, 6, 8, 9, 10
Maternal-Effect Genes and Body Plans	6, 7, 10, 11
Zygotic Genes and Segment Formation	7, 10
Cell-Cell Interactions in C. elegans	11

Vocabulary: Organization and Listing of Terms and Concepts

Structures and Substances

Zygote

Master regulatory genes

Binary switch genes

Vertebrates, invertebrates

   myoD, etc.

   myoblasts

   myogenin

Drosophila

   eyeless

   Small eye

   maternal effect genes

      anterior

      posterior

   zygotic genes

segmentation genes

   gap genes

   pair-rule genes

   segment polarity genes

selector genes

homeotic genes

   homeobox

   homeodomain

      Hox gene clusters

Arabidopsis

   MADS box proteins

Caenorhabditis elegans

   male, hermaphrodite

   lin, let, etc.

   programmed cell death

   ced genes

      vulva

Chapter 21 Developmental Genetics

Dolly

Callus

    myogenin

Molecular gradients

    anterior-posterior axis

    maternal-effect genes

Maternal cytoplasm

Blastoderm

Processes/Methods

Development

    cytoplasmic localization

    variable gene activity

    determination

        selective expression

        regulatory events

        patterns of gene activity

            cascades

            muscle cell development

    multistep

    progressive restriction

differentiation

    genetic and morphological

changes

cell-cell interaction

    intercellular communication

    vulval formation

Analyses

    enucleation of oocytes

    serial nuclear transfers

    C. elegans

    Drosophila

        oogenesis

        syncytial cellular blastoderm

        imaginal disks

        fate maps

        compartments

        metamorphosis

        molecular gradients

            anterior-posterior

            dorsal-ventral

        segmentation

        selector genes

            homeotic mutants

            positional cues

Concepts

Development (F21.1)

    genomic equivalence

    variable gene activity (F21.1)

    eukaryotes

        determination

        differentiation

        cell-cell interaction

Chapter 21 Developmental Genetics

totipotent

open strategy

genomic equivalence

progressive restriction

switch points, binary

homology

multistep

cascades

transcriptional events

different cells

different times

cell and tissue interactions

cytoplasmic localization

Maternal influences

anterior-posterior gradient

positional information

Cytoplasmic influences

Homeotic genes

Homeodomains

F21.1 Illustration of the relationship between determination and differentiation. Determination sets the program which will later be revealed by differentiation. The variable gene activity hypothesis suggests that different sets of genes are transcriptionally active in differentiated cells.

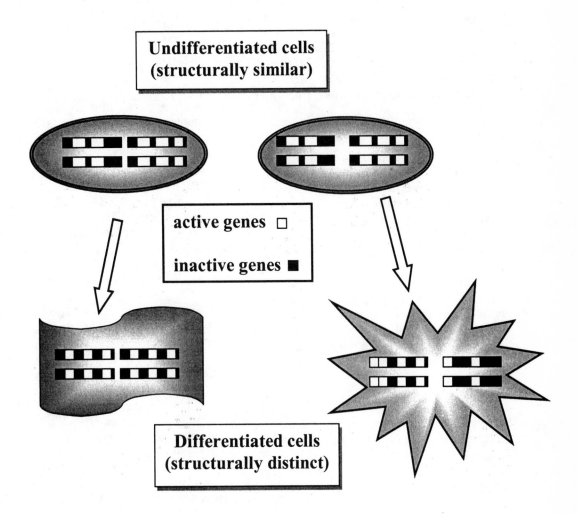

Solutions to Problems and Discussion Questions

1. *Determination* refers to early developmental and regulatory events which set eventual patterns of gene activity. Determination is not the end result of the regulatory activity, rather, it is the process by which the developmental fate of a particular cell type is fixed. *Differentiation* on the other hand follows determination and is the manifestation, in terms of genetic, physiological, and morphological changes, of the determined state.

2. Many of the appendages of the head, including the mouth parts and the antennae are evolutionary derivatives of ancestral leg structures. In *spineless aristapedia* the distal portion of the antenna is replaced by its ancestral counterpart, the distal portion of the leg (tarsal segments). Because the replacement of the arista (end of the antenna) can occur by a mutation in a single gene, one would consider that one "selector" gene distinguishes aristal from tarsal structures. Notice that a "one-step" change is involved in the interchange of leg and antennal structures.

3. Actinomycin D is useful in determining the involvement of transcription on molecular and developmental processes. Because maternal RNAs are present in the fertilized sea urchin egg (as is the case with many egg types) a considerable amount of development can occur without transcription. Because gastrulation is inhibited by *prior* treatment with actinomycin D, it would appear that earlier gene products are necessary for the initiation and/or continuation of gastrulation. Clearly a critical period (6th to 11th hours of development) exists for gastrulation in which gene activity is required.

4. There are several somewhat indirect methods for determining transcriptional activity of a given gene in different cell types. First, if protein products of a given gene are present in different cell types, it can be assumed that the responsible gene is being transcribed. Second, if one is able to actually observe, microscopically, gene activity, as is the case in some specialized chromosomes (polytene chromosomes), gene activity can be inferred by the presence of localized chromosomal puffs.

A more direct and common practice to assess transcription of particular genes is to use labeled probes. If a labeled probe can be obtained which contains base sequences that are complementary to the transcribed RNA, then such probes will hybridize to that RNA if present in different tissues. This technique is called *in situ* hybridization and is a powerful tool in the study of gene activity during development.

5. There are a variety of approaches to determine the level of control of a particular gene. First, one may detemine whether levels of hnRNA are consistent among various cell types of interest. This is often accomplished by either direct isolation of the RNA and assessment by northern blotting or by use of *in situ* hybridization. If the hnRNA pools for a given gene are consistent in various cell types, then transcriptional control can be eliminated as a possibility. Support for translational control can be achieved directly by determining, in different cell types, the presence of a variety of mRNA species with common sequences. This can be accomplished only in cases where sufficient knowledge exists for specific mRNA trapping or labeling. Clues as to translational control *via* alternative splicing can sometimes be achieved by examining the amino acid sequence of proteins. Similarities in certain structural/functional motifs may indicate alternative RNA processing.

6. Because in *ftz/ftz* embryos, the engrailed product is absent and in *en/en* embryos *ftz* expression is normal, one can conclude that the *ftz* gene product regulates, either directly or indirectly, *en*. Because the *ftz* gene is expressed normally in *en/en* embryos, the product of the *engrailed* gene does not regulate expression of *ftz*.

7. A *homeotic mutant* alters the identity of a segment or field within a segment as if it were another segment or field. While homologies do exist between widely diverse groups, some differences in function have evolved. It is likely that functional overlap would not occur, however, only an actual experiment would answer the question.

8. The fact that nuclei from almost any source remain transcriptionally and translationally active substantiates the fact that the genetic code and the ancillary processes of transcription and translation are compatible throughout the animal and plant kingdoms. Because the egg represents an isolated, "closed" system which can be mechanically, environmentally, and to some extent biochemically manipulated, various conditions may be developed which allow one to study facets of gene regulation. For instance, the influence of transcriptional enhancers and suppressors may be studied along with factors which impact on translational and post-translational processes. Combinations of injected nuclei may reveal nuclear-nuclear interactions which could not normally be studied by other methods.

9. The egg is not an unorganized collection of molecules from which life springs after fertilization. It is a highly organized structure, "preformed" in the sense that maternal informational molecules are oriented to provide an anterior-posterior and dorsal-ventral pattern from which nuclei receive positional cues. Such positional cues lead to the "determined" state, from which cells later reveal their adult form (differentiation).

Indeed, in *Drosophila* and many other organisms, embryonic fate maps may be constructed thereby attesting to the maternally-derived "prepattern" present in the egg. The egg therefore is preformed, not in the sense that a miniature individual resides, but in a molecular prepattern upon which development depends. However, work by Spemann, Briggs and King, and Gurdon, indicates that there is plasticity in the programming of nuclei and that even nuclei from somewhat specialized cells often have the potential to direct the development of the entire adult individual. Such *totipotent* behavior of cells indicates that development arises as a result of a series of progressive steps in which cells acquire new structures and functions as development progresses.

10. The typical developmental sequence for axis and segment formation in *Drosophila* proceeds from the gap genes to the pair-rule genes to the segment polarity genes. The fact that *fushi-tarazu* (*ftz*) is affected by early (anterior-posterior determining genes) and gap genes indicates that *ftz* functions after those genes. That segment polarity genes are influenced by *ftz* indicates that *ftz* functions earlier, thus placing *ftz* in the pair-rule group of genes.

11. (a) Since *her-1*⁻ mutations cause males to develop into hermaphrodites, and *tra-1*⁻ causes hermaphrodites to develop into males, one may hypothesize that the *her-1*⁺ gene produces a product which suppresses hermaphrodite development, while the *tra-1*⁺ gene product is needed for hermaphrodite development. Information provided in part (b) supports this hypothesis. **(b)** If the *her-1*⁺ product acts as a negative regulator, then when the gene is mutant, suppression over *tra-1*⁺ is lost and hermaphroditism would be the result. This hypothesis fits the information provided. The double mutant should be male because even though there is no suppression from *her-1*⁻, there is no *tra-1*⁺ product to support hermaphrodite development.

Chapter 22: Genetics and Cancer

Concept Areas	Corresponding Problems
Cell Cycles and Cancer	2, 3, 4
Genes and Cancer	1, 5, 9, 12, 13
Tumor Suppressor Genes	6, 7, 9, 12, 13
Oncogenes	8, 9
Genomic Changes and Cancer	9, 12, 13
Cancer and Environmental Agents	10, 11

Vocabulary: Organization and Listing of Terms and Concepts

Structures and Substances

Cancer susceptibility genes

Carcinogen

Saccharomyces cerevisiae

Schizosaccharomyces pombe

   G1, G2, S, G0

   checkpoints

  protein kinases

    cyclin-dependent kinase

   CDK1, *cdc2*

  cyclins (A, B, C, D1, D2, E)

CDK1/cyclin B

G1 checkpoint, M checkpoint

Tumor suppressor genes

  pRb

  RB, E2F

  proto-oncogenes (*c-onc*)

oncogenes (onc)

  v-onc, c-src, other oncogenes, *ras,*

  BRCA1 (chromosome 17)

  BRCA2 (chromosome 13)

    dominant genes

  RAD51

  MLH1, MSH2

  APC

  c-myc

  FCC

   zinc finger domains

   Philadelphia chromosome (CML)

Sarcoma

  retrovirus, Rous sarcoma virus

   reverse transcriptase

216

provirus

    acute transforming virus

    nonacute (nondefective) virus

    p53, "guardian of the genome"

      p21

    hepatitis B virus (HBV)

      gatekeeper genes

Extracellular matrix

    metalloproteinases

    TIMP

Processes/Methods

Metastasis

Cell cycle control

CDK/Cyclin complex phosphorylation

Cancer

    chromosomal changes

      loss, rearrangement, insertions

      chronic myeloid leukemia

      Philadelphia chromosome

    hybrid genes, hybrid proteins

    lymphoma

Genomic instability

    microsatellite DNA

    aneuploidy

Environmental factors

    hepatocellular carcinoma

    hepatitis B virus

      ionizing radiation

      chemicals

      diet, drugs

      ultraviolet light

Cell cycle control, "start"

    S phase control point

Genetic predisposition

    retinoblastoma

    autosomal dominant

      90% "penetrant"

        familial, sporatic

      phosphorylation

      dephosphorylation

Wilms tumor

    autosomal dominant

    tissue specific regulator

Apoptosis

    colon cancer

    hereditary nonpolyposis colorectal cancer (HNPCC)

    familial adenomatous polyposis (FAP)

    APC, p53, sequential aspects

Chapter 22 Genetics and Cancer

Concepts

Cellular basis of cancer

    genetic involvement

    somatic *versus* germline

Genetic influences, instability

    suppressor genes

    oncogenes, protooncogenes

Model for retinoblastoma control

Model for Wilms tumor control

Origin of oncogenes

Metastasis

Environmental involvement

———————

Solutions to Problems and Discussion Questions

1. Familial retinoblastoma is inherited as an autosomal dominant gene with 90% penetrance, that is, 90% of the individuals which inherit the gene will develop eye tumors. The gene usually expresses itself in youngsters. Because the husband's sister has RB, one of the husband's parents has the gene for RB and the husband has a 50:50 chance of inheriting that gene. However, because the husband is past the usual age of onset, it is quite likely that he was lucky and did not receive the RB gene. In that case, the chance that a child born to this couple having RB is no higher than the frequency of sporatic occurrence. However, because the gene is 90% penetrant, there is a chance that the husband has the gene but does not express it. The probability of that occurrence would be 0.50 (of inheriting the gene) X 0.10 (not expressing the gene) = 0.05. The chance of the husband then passing this non-expressed gene to his child would be again 0.5, so 0.50 X 0.05 = 0.025 for the child inheriting this gene. If the child inherits the RB gene, he/she has a 90% chance of expressing it. Therefore the overall probability of the child having RB (using this logic) would be 0.025 X 0.9 = 0.0225 or just over 2% (or about 1 in 50).

To test the presence of the RB gene in the husband, it is possible in some forms of RB to identify (by Southern blot) a defective or missing DNA segment. Otherwise, one might attempt to assay the RB product in cells to see if it is present and functional at normal levels.

2. Review Chapter 8 in the text and note that the following stages of the cell cycle are discussed: G_1, G_0, S, G_2. The G_1 stage begins after mitosis and is involved in the synthesis of many cytoplasmic elements. In the S phase DNA synthesis occurs. G_2 is a period of growth and preparation for mitosis. Most cell cycle time variation is caused by changes in the duration of G_1. G_0 is the non-dividing state.

3. The major regulatory points of the cell cycle include the following:

1. Late G_1 (G_1/S)
2. The border between G_2 and mitosis (G_2/M)
3. In mitosis (M)

4. Kinases regulate other proteins by adding phosphate groups. Cyclins bind to the kinases, switching them on and off. Several cyclins, including D and E, can move cells from G_1 to S. At the G_2/mitosis border a CDK1 (cyclin dependent kinase) combines with another cyclin (cyclin B). Phosphorylation occurs bringing about a series of changes in the nuclear membrane, cytoskeleton, and histone 1.

5. To say that a particular trait is inherited conveys the assumption that when a particular genetic circumstance is present, it will be revealed in the phenotype. For instance, albinism is inherited in such a way that individuals who are homozygous recessive, express albinism. When one discusses an inherited predisposition, one usually refers to situations where a particular phenotype is expressed in families in some consistent pattern. However, the phenotype may not always be expressed or may manifest itself in different ways. In retinoblastoma, the gene is inherited as an autosomal dominant and those that inherit the mutant RB allele are predisposed to develop eye tumors. However, approximately 10% of the people known to inherit the gene don't actually express it and in some cases expression involves only one eye rather than two.

6. A tumor suppressor gene is a gene that normally functions to suppress cell division. Since tumors and cancers represent a significant threat to survival and therefore Darwinian fitness, strong evolutionary forces would favor a variety of co-evolved and perhaps complex conditions in which mutations in these suppressor genes would be recessive.

Looking at it in another way, if a tumor suppressor gene makes a product that regulates the cell cycle favorably, cellular conditions have evolved in such a way that sufficient quantities of this gene product are made from just one gene (of the two present in each diploid individual) to provide normal function.

7. The dominantly inherited, *RB* gene, located on chromosome 13, encodes a 928 amino acid protein (pRb) which is present in all cell and tissue types in G_0 cells and those active in the cell cycle. When the RB protein (pRB) is dephosphorylated, it acts to suppress cell division by binding to and inactivating a transcription factor (E2F) thereby making pRb an important link between the cell cycle and gene transcription. If pRb is absent, E2F is not regulated and permanent expression of control genes occurs. This results in uncontrolled cell growth. Wilms tumor is caused by an autosomal dominant gene which also functions as a tumor suppressor. The *WT* gene is located on the short arm of chromosome 11 and encodes a protein with four contiguous zinc finger domains which are characteristic DNA binding proteins. The gene is activated only in mesenchymal cells of the fetal kidney and in the tumorous nephroblastoma cells. It is suggested that the *WT* gene product acts directly to turn off genes that sustain cell proliferation or turn on genes which differentiate mesenchymal cells into kidney cells. Both *pRb* and *WT* gene products are restricted to the nucleus but pRB does not bind to DNA, rather it appears to be a general regulator of cell division while the *WT* gene product is a cell or tissue-specific regulator.

8. Oncogenes are genes that induce or maintain uncontrolled cellular proliferation associated with cancer. They are mutant forms of proto-oncogenes which normally function to regulate cell division. They may be formed through point mutations, gene amplification, translocations, repositioning of regulatory sequences, *etc.*.

9. A translocation involving exchange of genetic material between chromosomes 9 and 22 is responsible for the generation of the "Philadelphia chromosome." Genetic mapping established that certain oncogenes were combined to form a hybrid gene that encodes a 200kd protein which has been implicated in the formation of chronic myelocytic leukemia.

10. Unfortunately, it is common to spend enormous amounts of money on dealing with diseases after they occur rather than concentrating on disease prevention. Too often pressure from special interest groups or lack of political stimulus retards advances in education and prevention. Obviously, it is less expensive, both in terms of human suffering and money, to seek preventive measures for as many diseases as possible. However, having gained some understanding of the mechanisms of disease, in this case cancer, it must also be stated that no matter what preventive measures are taken it will be impossible to completely eliminate disease from the human population. It is extremely important, however, that we increase efforts to educate and protect the human population from as many hazardous environmental agents as possible.

11. Any agent which causes damage to DNA is a potential carcinogen since cell cycle control is achieved by gene (DNA) products, known as proteins. Since cigarette smoke is known to contain an agent which changes DNA, in this case transversions, numerous modified gene products (including cell cycle controlling proteins) are likely to be produced. The fact that many cancer patients have such transversions in *p53* strongly suggests that cancer is caused by agents in cigarette smoke.

12. (a) The mRNA triplet for Gln is CAG(A). The mRNA triplet which specifies a stop is one of three: UAA, UAG, or UGA. The strand of DNA which codes for the CAG(A) would be the following: 3'-GTC(T)-5'. Therefore, if the G mutated to an A (transition), then the DNA strand would be 3'-ATC(T)-5' which would cause a UAG(A) triplet to be produced and this would cause the stop.

(b) Tumor suppression because loss-of-function causes predisposition to cancer.

(c) Some women may carry genes (perhaps mutant) which "spare" for the *BRCA1* gene product. Some women may have immune systems which recognize and destroy precancerous cells or they may have mutations in breast signal transduction genes so that cell division suppression occurs in the absence of *BRCA1*.

13. (a) Even though there are changes in the *BRCA1* gene, they don't always have physiological consequences. Such neutral polymorphisms make screening difficult, in that one can't always be certain that a mutation will cause problems for the patient.

(b) The polymorphism in *PM2* is probably a silent mutation because the third base of the codon is involved.

(c) The polymorphism in *PM3* is probably a neutral missense mutation because the first base is involved.

Chapter 23: Chromosome Genetics: Immunoglobins, Isochores and Chromatid Dynamics

Concept Areas   Corresponding Problems

Concept Areas	Corresponding Problems
Chromosome structure	7, 8
Somatic recombination	6, 11
Immune system	1, 2, 3, 4, 5
RAG-mediated recombination	6
Histone involvement in regulation	11
Gene distribution	8
Chromosome dynamics	9, 10, 12

Vocabulary: Organization and Listing of Terms and Concepts

Structures and Substances

Nucleosomes

Isochore genome organization

Immune system

   antibodies

   B cells

   antigens

   T cells

     receptors

   immunoglobins (Ig)

    IgG

     heavy chain (H)

      variable region (V_H)

       L-V (leader variable region)

       D (diversity)

J (joining)

C (constant)

constant region (C_H)

light chain (L)

variable region (V_L)

constant region (C_L)

kappa chains (chromosome 2)

  L-V (leader-variable)

  pseudogenes

  J (joining)

  C (constant)

lambda chains (chromosome 22)

antibody combining site

  germ line cells (V_L)

Chapter 23 Chromosome Genetics: Immunoglobins, Isochores and Chromatid Dynamics

T cell receptors (TCRs)

    alpha, beta

    gamma, delta

        V region (beta receptor gene)

        D region

        J region

        C region

Recombination activating genes

    RAG-1, *RAG*-2

      recombinational signal sequences (RSSs)

        heptamer (CACAGTG)

        nonamer (ACAAAACC)

          recombinase

            overhang

histone acetyltransferase (HAT)

histone deacetylase (HDAC)

telomere

heterochromatin

    Giesma, trypsin

subchromosomal domains

    GC/AT content

      satellites

    isochore

    CpG islands

    gene-rich and gene-poor regions

repetitive sequences

matrix attachment sites

chromosome bands

    C bands (centromeric regions)

    G bands (Giesma)

    R bands (reverse)

      T bands (telomeric)

    R' bands (not T bands)

gene "spaces"

    H families (isochores)

cohesin

    Smc1, Smc3, Scc1, Scc3

      SA1, SA2

      Eco1/Ctf7

separin (protease)

securin

anaphase promoting complex (APC)

Rec8

Processes/Methods

Chromosome remodeling

Bottom-up view of the chromosome

Molecular variability

    antibody

    receptors

Somatic recombination

Chapter 23 Chromosome Genetics: Immunoglobins, Isochores and Chromatid Dynamics

B cell maturation

    random joining by recombination

    break-nibble-add mechanism

    imprecise joining

V(D)J recombination

Allelic exclusion

RAG-mediated recombination

Histone acylation

    hyperacetylation

Gene distribution

Chromosome dynamics in mitosis

    sister chromatid cohesion

    sister chromatid separation

      anaphase transition

Chromosome dynamics in meiosis

Concepts

Chromosome structure and

remodeling

Molecular variability

Somatic recombination

B cell maturation

RAG-mediated recombination

Histone involvement in regulation

Gene distribution

Chromosome dynamics

    mitosis

Chromosome dynamics

    meiosis

Chapter 23 Chromosome Genetics: Immunoglobins, Isochores and Chromatid Dynamics

Solutions to Problems and Discussion Questions

1. Two major components of the immune system include the B cells and the T cells. B cells produce antibodies and T cells are involved in a variety of functions including recognizing and killing infected cells of the host. Antibodies made by B cells and receptors of T cells are encoded by a small family of genes.

2. Typically four disulfide bridges hold the two heavy and two light chains together in each antibody molecule. The disulfide bridges thereby contribute to the three dimensional structure and are dependent on cysteine residues.

3. V_L = variable region of the light chain, C_H = constant region of the heavy chain, IgG = an immunoglobin class which represents approximately 80% of the antibodies in the blood. J = genes that specify a portion of the V region which includes a portion of the hypervariable region. D = a region between V and J in the heavy immunoglobin chain.

4. The number of combinations is determined by a simple multiplication of the number of genes in each class: V X D X J X C. Thus in this case the answer would be 10 V X 30 D X 50 J X 3 C = 45,000.

5. The total number of combinations is determined by simple multiplication. In this case, for the heavy chain 5 V X 10 D X 20 J = 1000, and for the light chain 10 V X 100 J = 1000. The final total would be 1000 X 1000 = 10^6.

6. RAG proteins are involved in the first step of V(D)J recombination wherein they bind to RSS (recombination signal sequences). They produce double-stranded breaks between the RSS and coding DNA. When rejoining occurs, non-complementary base pairing may occur followed by trimming and filling. These processes create additional molecular diversity.

7. Notice that there is a higher percentage of the human genome contained in the H1-H3 isochores compared to yeast. Since the L isochore is relatively A-T rich we can conclude that there is a higher percentage of A-T in the yeast genome.

8. In the human genome, there are very few genes in low GC isochores (such as L1), and much higher numbers of genes in GC-rich H3 isochores. The distribution of genes is therefore quite non-uniform, with a small fraction of the genome containing the bulk of the genes.

9. Degradation of securin occurs at the anaphase transition and is initiated by the anaphase promoting complex (APC) which is controlled by the spindle assembly checkpoint. If a mutation or mutations lowered or eliminated function of APC, securin would remain intact and one would likely see a delay in the progression of the metaphase/anaphase transition.

10. Mitosis occurs with one round of DNA synthesis followed by one round of cell division, whereas, in meiosis one round of DNA synthesis is followed by two rounds of cell division. In addition to pairing of homologous chromosome during meiosis I, additional molecular events are unique. In meiotic yeast cells, the Scc1 component of cohesin is replaced by a meiosis-specific subunit called Rec8 which is broken down by separin.

11. Chromatin structure is intimately related to the expression and recombination of immunoglobin genes. N-terminal tails of histones are sites of acetylation/deacetylation, and V-J joining is known to be influenced by deacetylase inhibitors. Mutations in histone genes may alter histone acetylation and/or deacetylation associations and thereby influence V-J joining.

12. It is likely to have little direct effect on G1 because sister chromatids are not present as yet. However, overexpression of separin after S phase would probably cause a destruction of securins. This would result in premature chromatid separation and aneuploidy.

Chapter 24: Quantitative Genetics

Concept Areas	Corresponding Problems
Phenotypic Expression	8
Continuous Variation and Polygenes	1, 2, 3, 4, 5, 6, 7, 9, 16, 17, 18
Heritability	2, 10, 8, 11, 12, 13, 14, 15

Vocabulary: Organization and Listing of Terms and Concepts

Structures and Substances

DDT

QTLS

RFLP

Processes/Methods

Transmission genetics

Quantitative, polygenic inheritance

Discontinuous traits

Continuous traits

　multiple-factor hypothesis

　additive (cumulative, quantitative)

alleles

　nonadditive alleles

　polygenic

　$1/4^n$

　resistance to DDT (*Drosophila*)

Statistical analysis

　biometry

　descriptive summary

statistical inference

statistics

parameters

　mean

　　central tendency

　frequency distribution

　variance

　standard deviation

　standard error of the mean

Heritability

　broad-sense heritability

　　phenotypic variation

　　environmental variance

　　genetic variance

　　interaction

　narrow-sense heritability

　　additive variance

　　dominance variance

　　interactive variance

Chapter 24 Quantitative Genetics

Concepts

artificial selection

response (R)

selection differential (S)

twin studies

    monozygotic (identical) twins

    dizygotic (fraternal) twins

        concordant

        discordant

Transmission genetics

Quantitative inheritance

Heredity and environment

Mapping quantitative traits

    restriction fragment length polymorphism

Cosegregation

Heritability

    inbred lines

    heritability index (H^2)

        broad-sense heritability

        narrow-sense heritability

Solutions to Problems and Discussion Questions

1. In *discontinuous* variation the influences of each gene pair are not additive and more typical Mendelian ratios such as 9:3:3:1 and 3:1 result. In *continuous* variation, different gene pairs interact (usually additively) to produce a phenotype which is less "stepwise" in distribution. Inheritance of a quantitative nature follows a more continuous distribution.

2. (a) *Polygenes* are those genes which are involved in determining continuously varying or multiple factor traits.

(b) *Additive alleles* are those alleles which account for the hereditary influence on the phenotype in an additive way.

(c) The *multiple factor hypothesis* suggested that many factors or genes contribute to the phenotype in a cumulative or quantitative way.

(d) *Monozygotic twins* are derived from a single fertilized egg and are thus genetically identical to each other. They provide a method for determining the influence of genetics and environment on certain traits. *Dizygotic twins* arise from two eggs fertilized by two sperm cells. They have the same genetic relationship as siblings. The role of genetics and the role of the environment can be studied by comparing the expression of traits in monozygotic and dizygotic twins. The higher concordance value for monozygotic twins as compared to the value for dizygotic twins indicates a significant genetic component for a given trait.

(e) *Concordance* refers to the frequency with which both members of a twin pair express a given trait. *Discordance* refers to the frequency at which one twin expresses a trait while the other does not. A comparison of concordance (and discordance) frequencies can provide information on the genetic and/or environmental influence on a given trait.

(f) *Heritability* is a measure of the degree to which the phenotypic variation of a given trait is due to genetic factors. A high heritability indicates that genetic factors are major contributers to phenotypic variation while environmental factors have little impact.

3. If you add the numbers given for the ratio, you obtain the value of 16, which is indicative of a dihybrid cross. The distribution is that of a dihybrid cross with additive effects.

(a) Because a dihybrid result has been identified, there are two loci involved in the production of color. There are two alleles at each locus for a total of four alleles.

(b) Because the description of red, medium-red, *etc.*, gives us no indication of a *quantity* of color in any form of units, we would not be able to actually quantify a unit amount for each change in color. We can say that each gene (additive allele) provides an equal unit amount to the phenotype and the colors differ from each other in multiples of that unit amount.

(c) The genotypes are as follows:

1/16	= dark red	=	AABB
4/16	= medium-dark red	=	2AABb 2AaBB
6/16	= medium red	=	AAbb 4AaBb aaBB
4/16	= light red	=	2aaBb 2Aabb
1/16	= white	=	aabb

(d)

F_1 =   all light red

F_2 =   1/4 medium red
2/4 light red
1/4 white

229

4. (a) It is *possible* that two parents of moderate height can produce offspring that are much taller or shorter than either parent because segregation can produce a variety of gametes, therefore offspring as illustrated below:

$$rrSsTtuu \quad X \quad RrSsTtUu$$
(moderate)   (moderate)

Offspring from this cross can range from very tall *RrSSTTUu* (12 "tall" units) to very short *rrssttuu* (8 "small" units).

(b) If the individual with a minimum height, *rrssttuu*, is married to an individual of intermediate height *RrSsTtUu*, the offspring can be no taller than the height of the tallest parent. Notice that there is no way of having more than four uppercase alleles in the offspring.

5. As you read this question, notice that the strains are inbred, therefore homozygous, and that approximately 1/250 represent the shortest and tallest groups in the F_2 generation. See $\frac{1}{4}^n$ formula in the text.

(a, b) Referring to the text, see that where four gene pairs act additively, the proportion of one of the extreme phenotypes to the total number of offspring is 1/256 (add the numbers in each phenotypic class). The same may be said for the other extreme type. The extreme types in this problem are the 12cm and 36cm plants. From this observation one would suggest that there are four gene pairs involved.

(c) If there are four gene pairs, there are nine (2n+1) phenotypic categories and eight increments between these categories. Since there is a difference of 24cm between the extremes, 24cm/8 = 3cm for each increment (each of the additive alleles).

(d) A typical F_1 cross which produces a "typical" F_2 distribution would be where all gene pairs are heterozygous (*AaBbCcDd*), independently assorting, and additive. There are many possible sets of parents which would give an F_1 of this type.

The limitation is that each parent has genotypes which give a height of 24cm as stated in the problem. Because the parents are inbred, it is expected that they are fully homozygous. An example:

$$AABBccdd \quad X \quad aabbCCDD$$

(e) Since the aabbccdd genotype gives a height of 12cm and each upper-case allele adds 3cm to the height, there are many possibilities for an 18 cm plant:

AAbbccdd,

AaBbccdd,

aaBbCcdd, etc.

Any plant with seven upper-case letters will be 33cm tall:

AABBCCDd,

AABBCcDD,

AABbCCDD, for examples.

6. (a) There is a fairly continuous range of "quantitative" phenotypes in the F_2 and an F_1 which is between the phenotypes of the two parents; thererfore, one can conclude that some phenotypic blending is occurring which is probably the result of several gene pairs acting in an additive fashion. Because the extreme phenotypes (6cm and 30cm) each represent 1/64 of the total, it is likely that there are three gene pairs in this cross. Remember, trihybrid crosses which show independent assortment of genes have a denominator (4^3) of 64 in ratios. Also, the fact that there are seven categories of phenotypes, which, because of the relationship 2n+1 = 7, would give the number of gene pairs (n) of 3. The genotypes of the parents would be combinations of alleles which would produce a 6cm (*aabbcc*) tail and a 30cm (*AABBCC*) tail while the 18cm offspring would have a genotype of *AaBbCc*.

(b) A mating of an *AaBbCc* (for example) pig with the 6cm *aabbcc* pig would result in the following offspring:

Gametes (18cm tail)	Gamete (6cm tail)	Offspring
ABC		AaBbCc (18cm)
ABc		AaBbcc (14cm)
AbC		AabbCc (14cm)
Abc	abc	Aabbcc (10cm)
aBC		aaBbCc (14cm)
aBc		aaBbcc (10cm)
abC		aabbCc (10cm)
abc		aabbcc (6cm)

In this example, a 1:3:3:1 ratio is the result. However, had a different 18cm tailed-pig been selected, a different ratio would occur:

$$AABbcc \quad X \quad aabbcc$$

Gametes (18cm tail)	Gamete (6cm tail)	Offspring
ABc	abc	AaBbcc (14cm)
Abc		Aabbcc (10cm)

7. For height, notice that average differences between MZ twins reared together (1.7 cm) and those MZ twins reared apart (1.8 cm) are similar (meaning little environmental influence) and considerably less than differences of DZ twins (4.4 cm) or sibs (4.5) reared together. These data indicate that genetics plays a major role in determining height.

However, for weight, notice that MZ twins reared together have a much smaller (1.9 kg) difference than MZ twins reared apart, indicating that the environment has a considerable impact on weight. By comparing the weight differences of MZ twins reared apart with DZ twins and sibs reared together one can conclude that the environment has almost as much an influence on weight as genetics.

8. *Monozygotic twins* are derived from a single fertilized egg and are thus genetically identical to each other. They provide a method for determining the influence of genetics and environment on certain traits. *Dizygotic twins* arise from two eggs fertilized by two sperm cells. They have the same genetic relationship as siblings.

The role of genetics and the role of the environment can be studied by comparing the expression of traits in monozygotic and dizygotic twins. The higher concordance value for monozygotic twins as compared to the value for dizygotic twins indicates a significant genetic component for a given trait.

9. Many traits, especially those which we view as quantitative are likely to be determined by a polygenic mode with possible environmental influences. The following are some common examples: height, general body structure, skin color, and perhaps most common behavioral traits including intelligence.

10. At first glance, this problem looks as if it will be an arithmetic headache, however, the problem can be simplified.

(a) The mean is computed by adding the measurements of all of the individuals, then dividing by the number of individuals. In this case there are 760 corn plants. To keep from having to add 760 numbers, merely multiply each height group by the number of individuals in each group. Add all the products then divide by *n* (760). This gives a value for the mean of 140 cm.

(b) For the variance, use the formula given below (as in the text):

$$s^2 = V = n\Sigma f(x^2) - (\Sigma fx)^2 / n(n - 1)$$

To simplify the calculations, determine the square of each height group (100 cm for example) then multiply the value by the number in each group.

For the first group (100 cm) we would have:

100   X   100   X   20 = 200000

The rest of the groups are as follows:

110 X 110 X 60   = 726000
120 X 120 X 90   = 1296000
130 X 130 X 130 = 2197000
140 X 140 X 180 = 3528000
150 X 150 X 120 = 2700000
160 X 160 X 70   = 1792000
170 X 170 X 50   = 1445000
180 X 180 X 40   = 1296000

= 15180000

Now, the mean squared, multiplied by n is as follows:

140 X 140 X 760 = 14896000

Completing the calculations gives the following: (15180000 - 14896000)/759
= 284000/759
= 374.18

(c) The *standard deviation* is the square root of the variance or 19.34.

(d) The *standard error* of the mean is the standard deviation divided by the square root

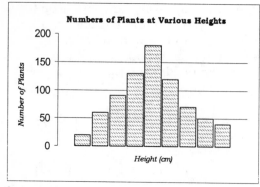

of *n*, or about 0.70.

The plot approximates a normal distribution. Variation is continuous.

11. For a trait that is quantitatively measured, the relative importance of genetic *versus* environmental factors may be formally assessed by examining the heritability index (H^2 or broad heritability). In animal and plant breeding, a measure of potential response to selection based on additive variance and dominance variance is termed narrow heritability (h^2). A relatively high narrow heritability is a prediction of the impact selection may have in altering an initial randomly breeding population.

12. The formula for estimating heritability is

$$H^2 = V_G/V_P$$

where V_G and V_P are the genetic and phenotypic components of variation, respectively. The main issue in this question is obtaining some estimate of two components of phenotypic variation: genetic and environmental. V_P is the combination of genetic and environmental variance. Because the two parental strains are inbred, they are assumed to be homozygous and the variance of 4.2 and 3.8 considered to be the result of environmental influences. The average of these two values is 4.0. The F_1 is also genetically homogeneous and gives us an additional estimation of the environmental factors.

By averaging with the parents

[(4.0 + 5.6)/2 = 4.8]

we obtain a relatively good idea of environmental impact on the phenotype. The phenotypic variance in the F_2 is the sum of the genetic (V_G) and environmental (V_E) components. We have estimated the environmental input as 4.8, so 10.3 (V_P) minus 4.8, gives us an estimate of (V_G) which is 5.5. Heritability then becomes 5.5/10.3 or 0.53. This value, when viewed in percentage form indicates that about 53% of the variation in plant height is due to genetic influences.

13. (a) For Vitamin A

$h_A^2 = V_A/V_P = V_A/(V_E + V_A + V_D) = 0.097$

For Cholesterol $h_A^2 = 0.223$

(b) Cholesterol content should be influenced to a greater extent by selection.

14. $h^2 = (7.5 - 8.5/6.0 - 8.5) = 0.4$

Selection will have little relative influence on olfactory learning in *Drosophila*.

15. $h^2 = 0.3 = (M_2 - 60/80 - 60)$

$M_2 = 66$ grams

16. Chromosome 2 seems to confer considerable resistance to the insecticide, somewhat in the heterozygous state and more in the homozygous state. Thus, some partial dominance is occurring.

17. (a) The most direct explanation would involve two gene pairs, with each additive gene contributing about 1.2mm to the phenotype. **(b)** The fit to this backcross supports the original hypothesis. **(c)** These data do not support the simple hypothesis provided in part (a). **(d)** With these data, one can see no distinct phenotypic classes suggesting that the environment may play a role in eye development or that there are more genes involved.

18. The best way to approach this problem is to first determine the number of gene pairs involved. Notice that all the F_1 plants are uniform and in the middle of the extremes of 3" and 15", therefore the parents must each be homozygous and at the extremes. Notice also that there are 13 classes in the F_2 so there must be six gene pairs. See the text for an explanation of the $2n+1$ formula and the examples in Figure 5-4. **(a)** There are two ways to answer this section, a hard way and an easy way. The hard way would to take a big sheet of paper, make the cross (*AaBbCcDdEeFf*   X   *AaBbCcDdEeFf*), collect the genotypes, and calculate the ratios.

This method would be very laborious and error-prone. The easy way would be to re-read the material on the binomial expansion and note the pattern preceding each expression. Notice that all numbers other than the 1's are equal to the sum of the two numbers directly above them. By enlarging the numbers in Figure 5-4 to include 6 gene pairs, you can arrive at the 13 classes and their frequencies:

3" =	1	4" =	12	5" =	66
6" =	220	7" =	495	8" =	792
9" =	924	10" =	792	11" =	495
12" =	220	13" =	66	14" =	12
15" =	1				

To check your calculations, be certain that your frequencies total 4096. You will also notice an additional shortcut in that since the distribution is symmetrical, you need only calculate to the center and the remainder will be in the reverse order. **(b)** To determine the outcome of a cross of the F_1 plants in the test cross, apply the formula which allows you to calculate any set of components: $n!/(s!t!)$ where n = total number of events (6), s = number of events of outcome a and t = number of events of outcome b. For example, to determine how many 6" plants would be recovered from the cross *AaBbCcDdEeFf*   X   *aabbccddeeff* we are really asking in how many will have three additive alleles (upper case) and three non-additive alleles (lower case).

$$6!/(3!3!) = 20$$

Applying this formula throughout gives the following frequencies.

3" =	1	4" =	6	5" =	15
6" =	20	7" =	15	8" =	6
9" =	1				

And the total is 64. You can check your logic by considering that there should be only 1/64 with no additive alleles (3") and 1/64 with all additive alleles (9").

Chapter 25: Population Genetics

<u>Concept Areas</u>	<u>Corresponding Problems</u>
Populations and Gene Pools	2
Calculating Allele Frequencies	1, 2, 11, 18, 20
The Hardy-Weinberg Law	3, 4, 6, 9, 10, 17
Extensions of the Hardy-Weinberg Law	21, 22
Using the Hardy-Weinberg Law	5, 7, 8, 9, 11
Factors that Alter Allele Frequencies	12, 13, 14, 15, 16, 19, 21

Vocabulary: Organization and Listing of Terms and Concepts

Historical

Charles Darwin

The Origin of Species - 1859

Alfred Russell Wallace

Structures and Substances

Population

Species

Gene pool

HIV-1

CC-CKR-5

CCR5

CCR5-Δ32

HLA-A

HLA-B

major histocompatibility complex (MHC)

I locus

acetylcholinesterase (ACE), Ace^R

chlorpyritos

aspartate aminotransferase 1

cystic fibrosis transmembrane

conductance regulator (CFTR)

FY-NULL

MN blood groups

Processes/Methods

Natural selection

fitness

Gene frequencies

mutation

migration

selection

random genetic drift

Chapter 25 Population Genetics

Symbolism

p, q

$p + q = 1$

$p^2 + 2pq + q^2 = 1$

Multiple alleles

$p + q + r = 1$

$p^2 + 2pq + 2pr + q^2 + 2qr + r^2 = 1$

Heterozygote frequency

$\sqrt{q^2}$

$p = 1 - q$

$2pq$

Demonstrating equilibrium

expected frequencies

observed frequencies

selection coefficient (s)

directional selection

stabilizing selection

disruptive selection

Drosophila

bristle number

Changes in gene frequencies

mutation (generates variability)

recessive

dominant

achondroplasia

migration

genetic drift

small populations

population size

Drosophila

forked bristles

isolated subpopulations

achromatopsia

Dunkers

ABO, MN

inbreeding and heterosis

assortative

inbreeding

self-fertilization

consanguineous marriages

coefficient of inbreeding

inbreeding depression

hybrid vigor

dominance theory

overdominance

Concepts

Population genetics

Gene pool

gene (allelic) frequencies (F25.1)

Population

Hardy-Weinberg Law

Hardy-Weinberg assumptions

    infinitely large

    no drift

    random mating

no selection

no mutation

no migration

Genetic equilibrium

    genetic variability

Inbreeding and hybrid vigor (F25.2)

Fitness

F25.1 Simple illustration of the relationships among populations, individuals, alleles, and allelic frequencies (p and q).

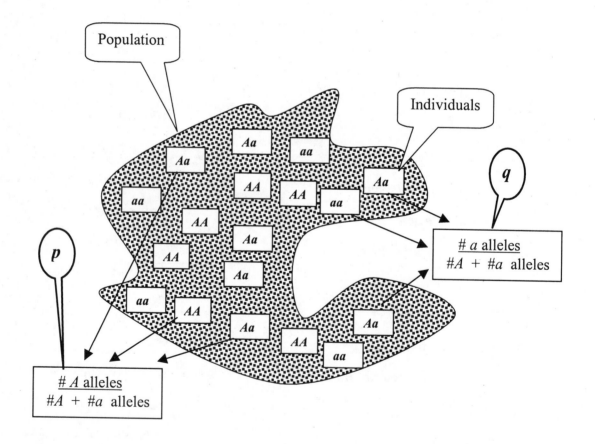

F25.2 Diagram of the relationships among inbreeding, heterosis, and homozygosity. Note that as inbreeding occurs, heterosis decreases while homozygosity increases.

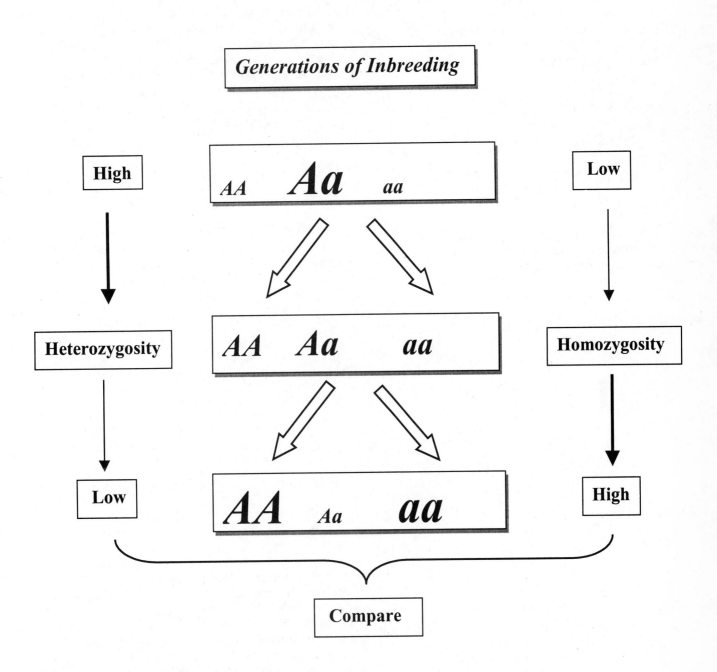

Chapter 25 Population Genetics

Solutions to Problems and Discussion Questions

1. Because the alleles follow a dominant/recessive mode, one can use the equation $\sqrt{q^2}$ to calculate q from which all other aspects of the answer depend. The frequency of aa types is determined by dividing 37 (number of non-tasters) by the total number of individuals (125).

q^2 = 37/125 = .296

q = .544

p = 1 - q

p = .456

The frequencies of the genotypes are determined by applying the formula $p^2 + 2pq + q^2$ as follows:

Frequency of AA = p^2

    = $(.456)^2$

    = .208 or 20.8%

Frequency of Aa = $2pq$

    = 2(.456)(.544)

    = .496 or 49.6%

Frequency of aa = q^2

    = $(.544)^2$

    = .296 or 29.6%

When completing such a set of calculations it is a good practice to add the final percentages to be certain that they total 100%.

2. Understanding the Hardy-Weinberg equilibrium allows us to state that if a population is equilibrium, the genotypic frequencies will not shift from one generation to the next unless there are factors such as selection, migration, *etc.* which alter gene frequencies. Since none of these factors are stated in the problem, we need only to determine whether the initial population is in equilibrium. Calculate p and q, then apply the equation $p^2 + 2pq + q^2$ to determine genotypic frequencies in the next generation.

p = frequency of A

    = 0.2 + .3

    = 0.5

q = 1 - p = 0.5

Frequency of AA = p^2

    = $(.5)^2$

    = .25 or 25%

Frequency of Aa = $2pq$

    = 2(.5)(.5)

    = .5 or 50%

Frequency of aa = q^2

    = $(.5)^2$

    = .25 or 25%

The initial population was not in equilibrium, however, after one generation of mating under the Hardy-Weinberg assumptions, the population is in equilibrium and will continue to be so (and not change) until one or more of the Hardy-Weinberg assumptions is not met. Note that *equilibrium* does not necessarily mean p and q equal 0.5.

238

3. For each of these values, one merely takes the square root to determine q, then one computes p, then one "plugs" the values into the $2pq$ expression.

(a) $q = .08$; $2pq$ $= 2(.92)(.08)$

$= .1472$ or 14.72%

(b) $q = .009$; $2pq$ $= 2(.991)(.009)$

$= .01784$ or 1.78%

(c) $q = .3$; $2pq$ $= 2(.7)(.3)$

$= .42$ or 42%

(d) $q = .1$; $2pq$ $= 2(.9)(.1)$

$= .18$ or 18%

(e) $q = .316$; $2pq$ $= 2(.684)(.316)$

$= .4323$ or 43.23%

(depending how one rounds off the decimals, slightly different answers will occur)

4. In order for the Hardy-Weinberg equations to apply, the population must be in equilibrium.

5. If one has the frequency of individuals with the dominant phenotype, the remainder have the recessive phenotype (q^2). With q^2 one can calculate q and from this value one can arrive at p. Applying the expression $p^2 + 2pq + q^2$ will allow a solution to the question.

6. (a) For the CCR5 analysis, first determine p and q. Since one has the frequencies of all the genotypes, one can add

$.6$ and $.351/2$ to provide p (= $.7755$);

q will be

$1 - .7755$ or $.2245$.

The equilibrium values will be as follows:

Frequency of $+/+ = p^2$ $= (.7755)^2$

$= .6014$ or 60.14%

Frequency of $+/\Delta32$ $= 2pq$

$= 2(.7755)(.2245)$

$= .3482$ or 34.82%

Frequency of $\Delta32 /\Delta32 = q^2 = (.2245)^2$

$= .0504$ or 5.04%

Comparing these equilibrium values with the observed values strongly suggests that the observed values are drawn from a population in equilibrium.

(b) For the AS analysis, first determine p and q. Since one has the frequencies of all the genotypes, one can add

$.756$ and $.242/2$ to provide p (= $.877$);

q will be

$1 - .877$ or $.123$.

The equilibrium values will be as follows:

Frequency of *AA* $= p^2 \quad = (.877)^2$

$= .7691$ or 76.91%

Frequency of *AS* $= 2pq \quad = 2(.877)(.123)$

$= .2157$ or 21.57%

Frequency of *SS* $= q^2 \quad = (.123)^2$

$= .0151$ or 1.51%

Comparing these equilibrium values with the observed values suggests that the observed values may be drawn from a population which is not in equilibrium. Notice that there are more heterozygotes than predicted, and fewer *SS* types. To test for a Hardy-Weinberg equilibrium, apply the chi-square test as follows.

$$\chi^2 = \frac{\Sigma(o\text{-}e)^2}{e}$$

$(75.6\text{-}76.9)^2 /76.9 +$

$(24.2 - 21.6)^2 /21.6 +$

$(0.2 - 1.51)^2 /1.5 \quad = \quad 1.47$

In calculating degrees of freedom in a test of gene frequencies, the "free variables" are reduced by an additional degree of freedom because one estimated a parameter (p or q) used in determining the expected values. Therefore, there is one degree of freedom even though there are three classes. Checking the χ^2 table with 1 degree of freedom gives a value of 3.84 at the 0.05 probability level.

Since the χ^2 value calculated here is smaller, the null hypothesis (the observed values fluctuate from the equilibrium values by chance and chance alone) should not be rejected. Thus the frequencies of *AA, AS, SS* sampled a population which is in equilibrium.

7. Given that $q^2 = .04$, then $q = .2$, $2pq = .32$, and $p^2 = .64$. Of those not expressing the trait, only a mating between heterozygotes can produce an offspring which expresses the trait, and then only at a frequency of $1/4$. The different types of matings possible (those without the trait) in the population, with their frequencies, are given below:

$AA \times AA$	$= .64 \times .64$	$= .4096$
$AA \times Aa$	$= .64 \times .32$	$= .2048$
$Aa \times AA$	$= .64 \times .32$	$= .2048$
$Aa \times Aa$	$= .32 \times .32$	$= .1024$
$Aa \times Aa$	$= .32 \times .32$	$= .1024$

Notice that of the matings of the individuals who do not express the trait, only the last two (about 20%) are capable of producing offspring with the trait. Therefore one would arrive at a final likelihood of $1/4 \times 20\%$ or 5% of the offspring with the trait.

8. The following formula calculates the frequency of an allele in the next generation for any selection scenario, given the frequencies of *a* and *A* in this generation and the fitness of all three genotypes.

$$q_{g+1} = [w_{Aa}p_g q_g + w_{aa}q_g^2]/[w_{AA}p_g^2 + w_{Aa}2p_g q_g + w_{aa}q_g^2]$$

where is the frequency of the *a* allele in the next generation, q_g is the frequency of the a allele in this generation, p_g is the frequency of the *A* allele in this generation, and each "w" represents the fitness of their respective genotypes.

(a)

$$q_{g+1} = [.9(.7)(.3)+.8(.3)^2/[1(.7)^2 +.9(2)(.7)(.3)+.8(.3)^2]$$

$q_{g+1} = \quad .278 \qquad p_{g+1} = \quad .722$

(b) q_{g+1} = .289 p_{g+1} = .711

(c) q_{g+1} = .298 p_{g+1} = .702

(d) q_{g+1} = .319 p_{g+1} = .681

9. The general equation for responding to this question is

$$q_n = q_0 / (1 + nq_0)$$

where n = the number of generations, q_0 = the initial gene frequency, and q_n = the new gene frequency.

(a)

$$q_n = q_0 / (1 + nq_0)$$

$$q_n = 0.5 / [1 + (1 \times 0.5)]$$

$$q_n = .33 \qquad p_n = .67$$

(b)

$$q_n = q_0 / (1 + nq_0)$$

$$q_n = 0.5 / [1 + (5 \times 0.5)]$$

$$q_n = .143 \qquad p_n = .857$$

(c)

$$q_n = q_0 / (1 + nq_0)$$

$$q_n = 0.5 / [1 + (10 \times 0.5)]$$

$$q_n = .083 \qquad p_n = .917$$

(d)

$$q_n = q_0 / (1 + nq_0)$$

$$q_n = 0.5 / [1 + (25 \times 0.5)]$$

$$q_n = .037 \qquad p_n = .963$$

(e)

$$q_n = q_0 / (1 + nq_0)$$

$$q_n = 0.5 / [1 + (100 \times 0.5)]$$

$$q_n = .0098 \qquad p_n = .9902$$

(f)

$$q_n = q_0 / (1 + nq_0)$$

$$q_n = 0.5 / [1 + (1000 \times 0.5)]$$

$$q_n = .00099 \qquad p_n = .99901$$

10. For this question, apply the equations

$$\Delta p = m(p_m - p)$$

and $p_1 = p + \Delta p$.

Substituting, gives: $p_1 = p + m(p_m - p)$.

(a) p_1 = 0.6 + 0.2(0.1 - 0.6) = 0.5

(b) p_1 = 0.2 + 0.3(0.7 - 0.2) = 0.35

(c) p_1 = 0.1 + 0.1(0.2 - 0.1) = 0.11

11. What one must do is predict the probability of one of the grandparents being heterozygous in this problem. Given the frequency of the disorder in the population as 1 in 10,000 individuals (0.0001), then q^2 = 0.0001, and $q = 0.01$. The frequency of heterozygosity is $2pq$ or approximately .02 as also stated in the problem. The probability for one of the grandparents to be heterozygous would therefore be 0.02 + 0.02 or 0.04 or 1/25.

If one of the grandparents is a carrier, then the probability of the offspring from a first-cousin mating being homozygous for the recessive gene is 1/16. Multiplying the two probabilities together gives 1/16 X 1/25 = 1/400.

Chapter 25 Population Genetics

Following the same analysis for the second-cousin mating gives 1/64 X 1/25 = 1/1600. Notice that the population at large has a frequency of homozygotes of 1/10,000, therefore one can easily see how inbreeding increases the likelihood of homozygosity.

12. *Inbreeding depression* refers to the reduction in fitness observed in populations which are inbred. With inbreeding comes an increase in the number of homozygous individuals (see F25.2 in this book) and a decrease in genetic variability. Genetic variability is necessary for a genetic response to environmental change. As deleterious genes become homozygous, more individuals are less fit in the population.

13. Because heterozygosity tends to mask expression of recessive genes which may be desirable in a domesticated animal or plant, inbreeding schemes will often be used to render strains homozygous so that such recessive genes can be expressed. In addition, assume that a particularly desirable trait occurs in a domesticated plant or animal. The best way to increase the frequency of individuals with that trait is by self-fertilization (not often possible) or by matings to blood relatives (inbreeding). In theory, one increases the likelihood of a gene "meeting itself" by various inbreeding schemes. There are disadvantages to increasing the degree of homozygosity by inbreeding. *Inbreeding depression* is a reduction in fitness often associated with an increase in homozygosity.

14. While inbreeding increases the frequency of homozygous individuals in a population, it does not change the *gene* frequencies. There will be fewer heterozygotes in the population to compensate for the additional homozygotes. See F25.2 in this book.

15. The quickest way to generate a homozygous line of an organism is to *self-fertilize* that organism. Because this is not always possible, brother-sister matings are often used.

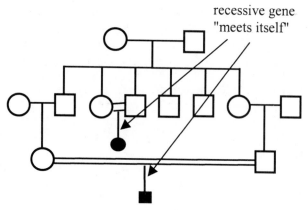

recessive gene "meets itself"

16. Given that the recessive gene a is present in the homozygous state (q^2) at a frequency of 0.0001, the value of q is 0.01 and $p = 0.99$.

(a) q is 0.01

(b) p $\quad = 1 - q$ or .99

(c) $2pq$ $\quad = 2(.01)(.99)$

$\quad\quad = 0.0198$ (or about 1/50)

(d) $2pq$ X $2pq$

$\quad\quad = 0.0198$ X 0.0198

$\quad\quad = 0.000392$ or about 1/255

242

17. The frequency of a gene is determined by a number of factors including the fitness it confers, mutation rate, and input from migration. There is no tendency for a gene to reach any artificial frequency such as 0.5. The distribution of a gene among individuals is determined by mating (population size, inbreeding, *etc.*) and environmental factors (selection, *etc.*). A population is in equilibrium when the distribution of genotypes occurs at or around the $p^2 + 2pq + q^2$ expression.

Equilibrium does not mean 25% *AA*, 50% *Aa*, and 25% *aa*. This confusion often stems from the 1:2:1 (or 3:1) ratio seen in Mendelian crosses.

18. Because three of the affected infants had affected parents, only two "new" genes, from mutation, enter into the problem. The gene is dominant, therefore each new case of achondroplasia arose from a single new mutation. There are 50,000 births, therefore 100,000 gametes (genes) involved. The frequency of mutation is therefore given as follows:

2/100,000

or 2×10^{-5}

19. The probability that the woman (with no family history of CF) is heterozygous is $2pq$ or $2(1/50)(49/50)$. The probability that the man is heterozygous is 2/3. The probability that a child with CF will be produced by two heterozygotes is 1/4. Therefore the overall probability of the couple producing a CF child is 98/2500 $\times$ 2/3 $\times$ 1/4.

20. (a) The gene is most likely recessive because all affected individuals have unaffected parents and the condition clearly runs in families. For the population, since q^2 = .002, then q = .045, p = .955, and $2(pq)$ = 0.086. For the community, since q^2 = .005, q = .07, p = .93, and $2(pq)$ = 0.13.

(b) The "founder effect" is probably operating here. Relatively small, local populations which are relatively isolated in a reproductive sense, tend to show differences in gene frequencies when compared to larger populations. In such small populations, homozygosity is increased as a gene has a higher probability of "meeting itself."

21. Given small populations and very similar environmental conditions, it is more likely that "sampling error" or genetic drift is operating. Under such conditions (small population sizes) large fluctuations in gene frequency are likely, regardless of selection pressures. Since the same gene is behaving differently under similar environmental conditions, selection is an unlikely explanation.

Chapter 26: Genetics and Evolution

Concept Areas	Corresponding Problems
Chromosomal Polymorphism	1
Models of Speciation	2
Measuring Genetic Variation	13
Formation of Species	10, 12
Molecular Techniques	3, 4, 5, 6, 7, 8, 9, 11, 13

Vocabulary: Organization and Listing of Terms and Concepts

Structures and Substances

Origin of Species (1859)

Allozyme

    alcohol dehydrogenase

    cystic fibrosis transmembrane

    conductance regulator (CFTR)

Cytochrome c

Polytene chromosome

Drosophila pseudoobscura

Fundulus heteroclitus (mummichog)

    lactate dehydrogenase

    ectotherm

Snapping shrimp

Mitochondrial DNA

Processes/Methods

Evolutionary divergence (F26.1)

    variation

    overpopulation

struggle for survival

differential survival

species formation

Genetic divergence

Genetic diversity

    heterozygosity

      protein polymorphism

      allozymes (F26.1)

        molecular phylogenetic trees

    amino acid sequence homology

      cytochrome c

    chromosomal polymorphism

      inversions

      translocations

DNA sequence polymorphism

    mitochondrial DNA

Chapter 26 Genetics and Evolution

Ecological diversity

   niche

Speciation

   stasis

   phyletic evolution (anagenesis)

   cladogenesis

   neutralist theory

   selectionist theory

   reproductive barriers

      physiological

      behavioral

      mechanical

   isolating mechanisms

      reproductive

         postzygotic

         prezygotic

Gel electrophoresis

   protein polymorphism

   nucleic acid sequence variation

UPGMA

Concepts

Niche

Evolutionary divergence

   molecular clock

   minimal mutational distance

   minimal genetic divergence

   rates of speciation

   divergence dendrograms

   evolutionary trees

Species formation (speciation)

Phylogenetic reconstruction

Sequence homology

   amino acid, nucleic acid

Sequence conservation

Parsimony

Maximum likelihood

Mutation and speciation

F26.1 The diagram below is meant to illustrate the meaning of the term allozyme. Notice that alleles A^1 and A^2 produce protein products which differ electrophoretically but accomplish the same function.

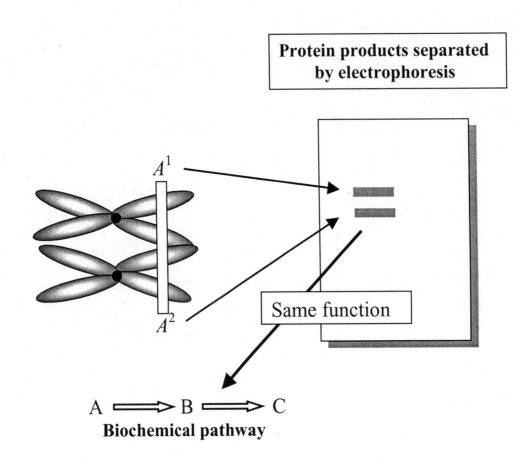

Chapter 26 Genetics and Evolution

Solutions to Problems and Discussion Questions

1. Assume that a chromosome in the "standard arrangement" undergoes an inversion (*pericentric* or *paracentric*). The following are possible consequences of such an inversion:

(a) change in gene order with possible introduction of position effects,

(b) breakage within a structural gene or other functional element,

(c) reduction in the recovery of crossover gametes in heterokaryotypes (those which carry an inversion as well as a standard homologue). While "c" may reduce the production of variation, the first two (a and b) may introduce variation.

At the population level, different populations with different inversion polymorphisms are genetically distinct.

2. During speciation, individuals or groups of potentially interbreeding organisms become genetically distinct from other members of the species. Members of different populations with substantial genetic divergence are, at first, not reproductively isolated from each other although gene flow may be restricted. The distinction between such groups is not absolute in that one group may blend with other groups of the species. Any process which favors changes in gene frequencies has the potential of generating substantial genetic differences.

Factors such as selection, migration, genetic drift, or even mutation, may be important in generating significant genetic change. One would certainly include geographic isolation as a major barrier to gene flow and thus an important process in such formation.

Natural selection occurs when there is non-random elimination of individuals from a population. Since such selection is a strong force in changing gene frequencies, it should also be considered as a significant factor in subspecies formation.

3. Because of degeneracy in the code, there are some nucleotide substitutions, especially in the third base, which do not change amino acids. In addition, if there is no change in the overall charge of the protein, it is likely that electrophoresis will not separate the variants. If a positively charged amino acid is replaced by an amino acid of like charge, then the overall charge on the protein is unchanged. The same may be said for other, negatively charged and neutral amino acid substitutions.

4. Construct a chart similar to the one below which indicates the number of base changes between each pair:

	H	C	G	O
H	-	-	-	-
C	1	-	-	-
G	3	2	-	-
O	7	6	4	-
B	12	11	9	10

Following the instructions given in the K/C text, develop the relationships in the following manner:

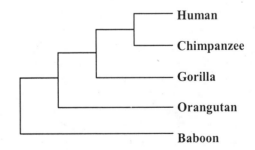

- Human
- Chimpanzee
- Gorilla
- Orangutan
- Baboon

5. All of the amino acid substitutions

(Ala - Gly, Val - Leu, Asp - Asn, Met - Leu)

require only one nucleotide change. The last change from

Pro (CC-) - Lys (AAA,G)

requires two changes (the minimal mutational distance).

6. Approach this problem by writing the possible codons for all the amino acids (except Arg and Asp which show no change) in the human cytochrome c chain. Then determine the minimum number of nucleotide substitutions required for each changed amino acid in the various organisms. Once listed, then count up the numbers for each organism: horse, 3; pig, 2; dog, 3; chicken, 3; bullfrog, 2; fungus, 6.

7. The classification of organisms into different species is based on evidence (morphological, genetic, ecological, *etc.*) that they are reproductively isolated. That is, there must be evidence that gene flow does not occur among the groups being called different species. Classifications above the species level (genus, family, *etc.*) are not based on such empirical data. Indeed, classification above the species level is somewhat arbitrary and based on traditions which extend far beyond DNA sequence information. In addition, recall that DNA sequence divergence is not always directly proportional to morphological, behavioral, or ecological divergence.

While the genus classifications provided in this problem seem to be invalid, other factors, well beyond simple DNA sequence comparison, must be considered in classification practices. As more information is gained on the meaning of DNA sequence differences (ΔT_m) in comparison to morphological factors, many phylogenetic relationships will be reconsidered and it is possible that adjustments will be needed in some classification schemes.

8. In looking at the figure, notice that the $\Delta T_{50}H$ value of 4.0 on the right could be used as a decision point such that any group which diverged above that line would be considered in the same genus while any group below would be in a different genus. Under this rule, one would have the chimpanzee, pygmy chimpanzee, human, gorilla and orangutan in the same genus. If one assumed that 3.7 is close enough to be considered above 4.0, given considerable experimental error, one could provide a scheme where the orangutan is not included with the chimpanzee, pygmy chimpanzee, human, and gorilla.

9. There are many sections of DNA in a eukaryotic genome which are not reflected in a protein product. Indeed, there are many sections of DNA which are not even transcribed and/or have no apparent physiological role. Such regions are more likely to tolerate nucleotide changes compared to those regions with a necessary physiological impact. Introns for example show sequence variation which is not reflected in a protein product. Exons on the other hand code for products which are usually involved in production of a phenotype and as such are subject to selection.

10. The text lists several cornerstones of the *neutral mutation theory:*

(a) the relatively uniform rate of amino acid substitution in different organisms (under different types of selection);

(b) there is no particular pattern to the substitutions indicating that selection is not eliminating some variations;

(c) the rate of mutation is relatively high and has remained relatively constant for millions of years even though environments have fluctuated greatly over that period of time;

(d) certain regions of molecules and certain functions of those molecules should logically be less likely to have amino acid substitutions influence the phenotype.

(e) the rate of amino acid substitution in some proteins is much too high to have been produced by selection. The *selectionists* suggest that even though amino acid substitutions *appear* to be neutral, it is more likely that their influence has just not been determined. In addition, they point out that many polymorphisms are clearly maintained in the population *by* selection. Thus, the issues listed above don't really challenge present views of genetic variation.

11. Like many other debates which surround the nature of evolution, it is important to see that debate is a natural component of scientific understanding. It is likely that some genes (like histones) will not tolerate nucleotide substitutions to a significant degree and the neutral mutation theory will not hold. However, there are other genes which produce quite variable products and provide support for the neutral mutation theory. Usually controversy is resolved as one dives deeper into the problem and seeks to define the variables and complexities of the process. It is controversy which stimulates a desire to seek answers.

12. Given the small range of *HLA* diversity, one might conclude that the Native Americans descended from a relatively small population either because of a small number who originally arrived or because of significant population crashes over time. Additionally, either the South American groups received an influx of genes from other populations or only small bands of North Americans survived.

13. The increased frequency of diabetics seems to correspond with an increase in sugar intake perhaps compounded by genotypes which are very efficient in sugar conversions. In the post 1970 period, the decline may be due to selection against the gene. Diabetic women pass fewer of the diabetic genes to the next generation.

Chapter 27: Conservation Genetics

Concept Areas	Corresponding Problems
Population Dynamics	1, 3
Genetic Assessment of Threatened Species	1
Management of Threatened Species	2, 5, 6, 7
Inbreeding and Drift in Small Populations	3, 4, 6, 8

Vocabulary: Organization and Listing of Terms and Concepts

Structures and Substances

Pacific yew (*Taxus brevifolia*)

Grey wolf (*Canis lupus*)

California condors

    (*Gymnogyps californianus*)

Russian wheat aphid (*Diuraphis noxia*)

Antarctic fur seal (*Arctocephalus sp.*)

Cheetah (*Acinonyx jubatus*)

North American brown bear
(*Ursus arctos*)

Peppered moth (*Biston betularia*)

Fruit fly (*Drosophila melanogaster*)

Black-footed ferret (*Mustela nigripes*)

Native rock grape (*Vitis rupestris*)

Domesticated species

Isozyme

DNA profile

    nuclear

    mitochondrial

    chloroplast

Harmonic mean

Gene bank

Core collection

Processes/Methods

Human population growth

Biodiversity

    human impact

Conservation genetics

Intraspecific diversity

    interpopulation

    intrapopulation

Interspecific diversity

Chapter 27 Conservation Genetics

Loss of genetic diversity

    reduced population size

    habitat loss

    population fragmentation

Detection of genetic diversity

    isozyme analysis

    RFLP

    PCR

    AFLP (amplified fragment length
        polymorphism)

    naturally rare

    newly rare

Absolute population size (N)

    effective populaton size (N_e)

    $N_e = 4(N_m N_f)/(N_m + N_f)$

    $N_e = 1/(1/t)(1/N_1 + 1/N_2 + 1/N_3 \ldots)$

Population bottleneck

Founder effect

Genetic drift

Inbreeding

Concepts

Vulnerable and endangered species

Genetic diversity

Loss of genetic diversity

    population size

Population dynamics

    bottleneck

    founder effect

    genetic drift

    inbreeding

    gene flow

    genetic erosion

Conservation strategies

Probability of fixation $p(A)$

Inbreeding coefficient

    $F = (2pq - H)/2pq$

    $H_t = (1-1/2N_e)^t H_o$

    inbreeding depression

Genetic load

    purging genetic load

Gene flow relates to migration

Genetic erosion (loss of diversity)

Ex situ and *In situ* conservation

    captive species

    gene banks

Population augmentation

    outbreeding depression

Chapter 27 Conservation Genetics

Solutions to Problems and Discussion Questions

1. (a) Apply the formula which computes the effective population size as the harmonic mean of the numbers in each generation:

$$N_e = 1/(1/t)(1/N_1 + 1/N_2 + 1/N_3 \ldots)$$

Substituting the values:

$$N_e = 1/(1/4)(1/47 + 1/17 + 1/20 + 1/35)$$

$$N_e = 25.21$$

(b) Apply the formula which computes the frequency of heterozygotes after t generations as a function of effective population size:

$$H_t = (1 - 1/2N_e)^t H_o$$

Substituting the values:

$$H_t = (1 - 1/2(25.21))^4 (.55)$$

$$H_t = .5073$$

(c) Apply the foumula which relates the inbreeding coefficient to the frequency of heterozygotes in a population:

$$F = (2pq - H)/2pq$$

$$F = (.55 - .5073)/.55$$

$$F = 0.0776$$

2. The frequency of the lethal gene in the captive population ($q^2 = 5/169$ and $q = .172$) is approximately double that in the gene pool as a whole ($q = 0.09$). Applying the formula

$$q_n = q_o/(1 + nq_o)$$

one can estimate that it would take 10 generations to reduce the lethal gene's frequency to .063 in the captive population with no intervention (random mating assumed). Since condors produce very few eggs per year, a more proactive approach seems justified.

First, if detailed records are kept of the breeding partners of the captive birds, then knowledge of heterozygotes should be available. Breeding programs could be established to restrict matings between those carrying the lethal gene. Such "kinship management" is often used in captive populations. If kinship records are not available, it is often possible to establish kinship using genetic markers such as DNA microsatellite polymorphisms. Using such markers, one can often identify mating partners and link them to their offspring.

By coupling knowledge of mating partners with the likelihood of producing a lethal genetic combination, selective matings can often be used to minimize the influence of a deleterious gene. In addition, such markers can be used to establish matings which optimize genetic mixing, thus reducing inbreeding depression.

3. Notice (in the text) that the probability of fixation through drift is the same as a gene's initial frequency. In this problem, the probability of A being fixed (and therefore a lost) is 0.75. The probability of B being fixed (and b being lost) is 0.8 and the probability of C being fixed (and c being lost) is 0.95. Therefore the probability that all the recessive alleles will be lost through genetic drift is

$$0.75 \times 0.80 \times 0.95 = 0.57$$

4. Both genetic drift and inbreeding tend to drive populations toward homozygosity. Genetic drift is more common when the effective breeding size of the population is low. When this condition prevails, inbreeding is also much more likely. They are different in that inbreeding can occur when certain population structures or behaviors favor matings between relatives, regardless of the effective size of the population. Inbreeding tends to increase the frequency of both homozygous classes at the expense of the heterozygotes. Genetic drift can lead to fixation of one allele or the other, thus producing a single homozygous class.

5. There are a number of dangers inherent in the management of such a small herd of endangered rhinos. Because of the small breeding pool, inbreeding depression is likely to lead to less fit individuals over time. To combat this problem, genetic markers (such as microsatellites) can be used to assess the general degree of relatedness and heterozygosity of each of the 16 rhinos. From such information, appropriate matings can be facilitated with would reduce inbreeding depression. However, additional efforts may be needed in this extreme case. It is possible to develop exchange programs whereby animals from other herds provide semen (either naturally or artificially) thereby reducing inbreeding. This practice can be successful if females are receptive and if no deleterious genes are brought into the population (outbreeding depression).

Population augmentation, where individuals are transplanted into a declining population, can be used to increase numbers and genetic diversity. However, as stated above, outbreeding depression accompanies this practice.

Sometimes drastic measures must be taken in extreme cases such as the black rhino. Dehorning is often practiced to remove the incentive for poaching and reduce lethal wounding due to fighting. This practice is only useful in areas void of dangerous predators.

6. Inbreeding depression, over time, reduces the level of heterozygosity, usually a selectively advantageous quality of a species. When homozygosity increases (through loss of heterozygosity) deleterious alleles are likely to become more of a load on a population. Outbreeding depression occurs when there is a reduction in fitness of progeny from genetically diverse individuals. It is usually attributed to offspring being less well-adapted to the local environmental conditions of the parents.

Even though forced outbreeding may be necessary to save a threatened species, where population numbers are low, it significantly and permanently changes the genetic make-up of the species.

7. Cloning of some highly threatened species may be the only way to save that species from extinction. However, the long-term disadvantages of cloning for this purpose are often considered self-defeating. With cloning one "short-circuits" normal processes (meiosis, gametic union, *etc.*) necessary to maintain genetic variation. With a loss of genetic variation comes difficulties with adaptation as environments change. It may be possible to identify certain conditions in which cloning would be useful to "save" a species, however, interbreeding provides benefits which allow a species to evolve. In addition, as members of a species become more uniform (through cloning) they are more likely to suffer more severe and widespread responses to disease and environmental stress.

8. Often, molecular assays of overall heterozygosity can indicate the degree of inbreeding and/or genetic drift. Refer to Figure 25.2 in this book and notice that as inbreeding (and genetic drift for that matter) occurs, the degree of heterozygosity decreases. An allele whose frequency is dictated by inbreeding and/or will not be uniquely influenced. That is, other alleles would be characterized by decreased heterozygosity as well. So, if the genome in general has a relatively high degree of heterozygosity, the gene is probably influenced by selection rather than inbreeding and/or genetic drift.

Sample Test Questions

(detailed explanations of answers follow in next section)

How to use this section:

The purpose of these *Sample Test Questions* is to present a slightly different style of question. Set aside several hours of study time, perhaps a week before each examination. Select two questions from each chapter covered on your upcoming test. Attempt to work selected questions, five or so per hour, under test conditions. **Write down your answers**, then, *after* you have finished the entire "test," check your answers.

If you are having difficulty, then you are weak in the concept areas listed for each question. If you have made mistakes, take comfort, there are many places to make mistakes on these problems. Some of the students who made the same mistakes are now practicing geneticists!

For the questions below from Chapter 1, there are few major concepts. This chapter serves as an introduction and to give you a general overview of the content of the text and its significance. Test questions are primarily oriented toward reading retention and study effort.

Ch.1 Ques.1 In 1859, Charles Darwin published *The Origin of Species* in which he presented ideas on the causes of organismic change through time. A primary conceptual gap existed which left his theory open to criticism. What was that conceptual gap?

Ch.1 Ques.2 Name the individual who, working with the garden pea in the mid 1850s, demonstrated quantitative patterns of heredity and developed a theory involving the behavior of hereditary factors.

Ch.1 Ques.3 What does the term "genetics" mean?

Ch.1 Ques.4 Name the substance which serves as the hereditary material in eukaryotes and prokaryotes. Is your answer the same for viruses?

Sample Test Questions

Ch.1 Ques.5 When examining chromosomes from a single nucleus of an individual cell it is often possible to match up chromosomes on the basis of overall size, centromere position, and sometimes other physical characteristics. Chromosomes which can be matched up or paired are called _____.

Ch.1 Ques.6 What is the difference between deoxyribonucleic acid and ribonucleic acid?

Ch.1 Ques.7 List three components of the genetic material, DNA.

> **Concepts: understanding of gene function, understanding DNA structure, function, mutation, variation**

Chs. 2, 3 Ques.8 The foundations of molecular genetics rest upon the assumption that a genetic material exists with the following properties:

    a. Autocatalytic (can replicate itself)

    b. Heterocatalytic (can direct form and function)

    c. Mutable

    d. Can exist in an infinite number of forms

(a) Provide a simple sketch which demonstrates the replication scheme of DNA.

(b) Briefly describe how DNA provides form and function.

(c) At the level of nucleotides, what characterizes mutant DNA?

(d) Why may we say that DNA can exist in an infinite number of forms?

Sample Test Questions

Ch. 2, 3 Ques.9 On the graph below draw C_{ot} curves for DNA from two genomes, one lacking repetitive DNA and the other containing repetitive DNA. Indicate the point on each curve at which the renaturation is half-complete. Label the horizontal and vertical axes accordingly. Explain the molecular basis for the different curves.

Ch. 2, 3 Ques.10 Given below is a single-stranded nucleotide sequence. Answer questions which refer to this sequence.

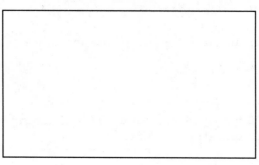

Sample Test Questions

a. In the circle at the bottom of this sequence, place a 5' or 3', whichever corresponds.

b. Is the above structure an RNA or DNA? State which_____.

c. Assume that a complementary strand is produced in which all the innermost phosphates of the adenine triphosphonucleoside precursors are labeled with ^{32}P. What bases would be labeled if the complementary strand is completely degraded with spleen diesterase (cleaves between the phosphate and the 5'carbon)?_____.

d. What bases would be labeled if the complementary strand was completely degraded with snake venom diesterase (cleaves between the phosphate and the 3' carbon)?_____.

> **Concepts: semiconservative replication, labeling, centrifugation, denaturation**

Ch.3 Ques.11 Assume that you were able to culture a strain of *E. coli* in medium containing either "normal" nitrogen or a heavy isotope of nitrogen (^{15}N). You grow the bacteria for a time in ^{15}N-containing medium which permits one complete replication of the bacterial chromosome. You extract the DNA, calling this extraction A. You continue to grow the bacterial culture in the ^{15}N DNA for a time which permits one more complete round of chromosome replication. You again extract the DNA, calling this extraction B. Assuming that non-labeled DNA has a density of 1.6 and that fully labeled DNA (that is with **all** the ^{14}N replaced with ^{15}N) has a density of 1.9, construct sedimentation profiles which reflect the expected densities of DNA from extractions A and B.

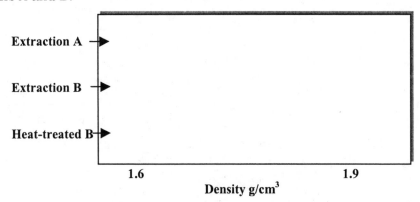

Knowing that heating DNA to 100° C. causes separation of complementary strands, use a broken line (- - -) to indicate the sedimentation profile of heat denaturation of extraction B DNA.

Sample Test Questions

> **Concepts: semiconservative replication, chromosome morphology, DNA structure, labeling, enzymatic digestion, 5', 3' orientations**

Ch. 3 Ques.12 Assume that you are microscopically examining mitotic metaphase cells of an organism with a $2n$ chromosome number of 2 (both telocentric). Assume also that the cell passed through one S phase labeling (innermost phosphate of dCTP radioactive) just prior to the period of observation.

(a) Draw this cell's chromosomes, and the autoradiographic pattern you would expect to see.

(b) Assuming that the A+T/G+C ratio of the DNA in this cell is 1.67 and this DNA is digested with snake venom diesterase (cleaves at the 3' position), what percentages of the total radioactivity would the following products have?

Adenine_____   Guanine_____
Thymine_____   Cytosine_____

> **Concepts: overall DNA replication, 5', 3' polarity restrictions, enzymology, priming**

Ch. 3 Ques.13 Drawn below is a diagram (not to scale) of DNA in the process of replication. Numbered arrows point to specific structures which you are to identify in the corresponding spaces below:

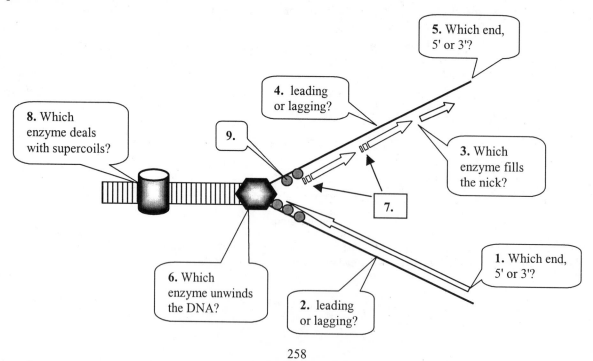

5. Which end, 5' or 3'?

4. leading or lagging?

8. Which enzyme deals with supercoils?

9.

3. Which enzyme fills the nick?

7.

1. Which end, 5' or 3'?

6. Which enzyme unwinds the DNA?

2. leading or lagging?

258

Sample Test Questions

1. _____ 2. _____ 3. _____

4. _____ 5. _____ 6. _____

7. _____ 8. _____ 9. _____

> **Concepts: transcription 5', 3' orientations, translation tRNA orientation, general process, rRNA in ribosomes**

Ch.5, 6 Ques.14 Below is a schematic of transcription and translation occurring simultaneously as described by Miller *et al.* (1970) in *E. coli*.

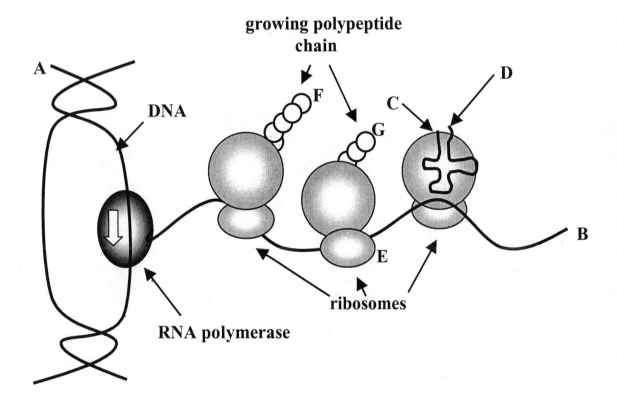

growing polypeptide
chain

DNA

RNA polymerase

ribosomes

Sample Test Questions

1. Is "A" at the 5' or 3' end (*state which*) of the DNA strand? _____

2. Is "B" at the 5' or 3' end (*state which*) of the RNA strand? _____

3. Is "C" at the 5' or 3' end (*state which*) of the RNA strand? _____

4. To what type of RNA does "D" point? _____

5. Would base sequences near letters "C" or "D" (state which) be expected to hold the amino acid? _____

6. What is the S value of the rRNA in the small subunit of the ribosome closest to letter "E"?_____

7. Is the amino acid nearest letter "F" the same type as the one nearest letter "G" (yes or no)? _____

Concepts: translation, coding, mutation

Chs. 6, 7 Ques.15 Assume that the following sequence of amino acids occurs in a protein starting from the "N" terminus (with asp) of a large polypeptide chain:

asp-glu-ile-leu-ser-thr-met-arg-tyr-try-phe-gly

Assume that gene X is responsible for synthesis of this gene. Answer the questions below.

a. Which amino acid(s) would you expect to change if gene X is altered by the mutagen, 2-amino purine, such that a transition mutation occurred which caused a change in the 9th base of the mRNA (counting from the 5' end of the coding region)?

b. Which amino acid(s) would you expect to change if gene X is altered by mutagen, such as acridine orange, such that a frameshift mutation occurred which caused an insertion of a base between bases 3 and 4 of the mRNA (counting from the 5' end of the coding region)?

c. Which amino acid(s) would you expect to change if gene X is altered by the mutagen, nitrous acid, such that a mutation occurred which caused a change in the 11th base of the mRNA (counting from the 5' end of the coding region)?

Sample Test Questions

Concepts: pathway analysis, Beadle and Tatum "set-up",
biochemical (nutritional) phenotype

Chs. 6, 7 Ques.16 Below is a set of experimental results relating the growth (+) of *Neurospora* on several media. Based on the information provided, present the biochemical pathway and the locations of the metabolic blocks.

Strain	Medium		
	MM	MM+A	MM+B
t409	-	+	+
t410	+	+	+
r3	-	-	+

Concepts: importance of primary structure, varieties of bonds "higher level" folding
structure/function relationships

Ch. 5 Ques.17 Drawn below is a hypothetical protein which contains areas where various types of bonds might be expected to occur. For each area a box is drawn and in that box is placed a number. In the corresponding spaces below, state which bond type (or interaction) is most likely illustrated **and** state how that particular type of bond (or interaction) is formed. *You may use a given bond type only once.*

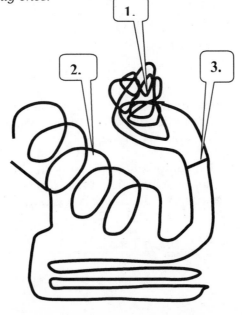

Sample Test Questions

1.

2.

3.

4. What type of amino acids tend to be located on the outside (water side) of the molecule?

> **Concepts: overlapping genes, differential hnRNA splicing**

Ch.5, 6 Ques.18 Some viruses as well as eukaryotes have evolved different mechanisms for obtaining more than one kind of protein from a single transcription unit (a transcription unit simply being a stretch of DNA that is transcribed into a single primary RNA transcript). Describe two different mechanisms.

> **Concepts: chromosome mechanics (mitosis, meiosis), symbolism, DNA content (cell cycles)**

Ch. 8 Ques. 19 The mosquito, *Culex pipiens*, has a diploid chromosome number of 6. Assume that one chromosome pair is metacentric, and the other two pairs are acrocentric.

(a) Draw chromosomal configurations which one would expect to see at the following stages: primary oocyte (metaphase I), secondary spermatocyte.

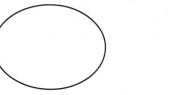

(b) Assuming that a G1 nucleus in *Culex* contains about 20 picograms (pg) of DNA, how much DNA would you expect in the following nuclei; Primary Spermatocyte, First Polar Body, Secondary Oocyte, Ootid in G1 phase.

(c) Assume that a female mosquito is heterozygous for the recessive gene *wavy bristles* (symbolized as *wb*) and this gene locus is on an acrocentric chromosome. Draw an expected mitotic metaphase with the appropriate genetic labeling pattern.

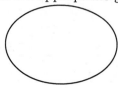

Sample Test Questions

Concepts: chromosome mechanics (mitosis, meiosis), symbolism, DNA content (cell cycles)

Ch. 8 Ques.20 In humans, chromosome #1 is large and metacentric, the X chromosome is medium in size and submetacentric (submedian), while the Y chromosome is small and acrocentric. Assume that you were microscopically examining human chromosomes at the stages given below.

(a) Illustrate (draw) the above-mentioned (#1, X and/or Y) chromosomes and/or pairs at the stages given (several different configurations may be applicable in some cases):

Metaphase I (Primary Oocyte):

First Polar Body:

Secondary Spermatocyte:

Secondary Oocyte:

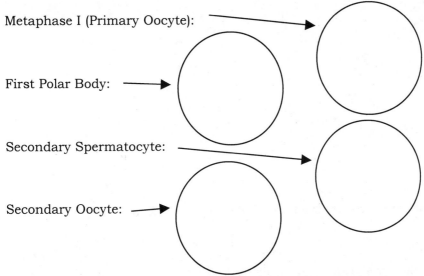

(b) The Rh blood group locus is on Chromosome #1. Individuals are *DD* or *Dd* if Rh⁺ and *dd* if Rh⁻. The locus for glucose-6-phosphate-dehydrogenase deficiency (*G6PD*) is located on the X chromosome. There are two alternatives at this locus, + and -. For each of the above cells, place genes (using symbolism given) on chromosomes if the female is heterozygous at both the *Rh* and *G6PD* loci. Do the same for the secondary spermatocyte (above) assuming that the male is Rh⁻, and + for the *G6PD* locus.

(c) Assume that the average DNA content per G1 nucleus in humans is 6.5 picograms. For the nuclei (including the entire chromosome complement for each nucleus) presented, give the expected DNA content:

Metaphase I (Primary Oocyte):_____   Secondary Spermatocyte:_____

First Polar Body:_____   Secondary Oocyte:_____

263

> **Concepts: chromosome morphology (telocentric, etc.), anaphase configurations chromosome mechanics, meiosis, mitosis, DNA content in cell cycles**

Ch.8 Ques.21 Assume that you are examining a cell under a microscope and you observe the following as the total chromosomal constituents of a nucleus. You know that $2n = 2$ in this organism, that all chromosomes are telocentric, and that each G1 cell nucleus contains 8 picograms of DNA.

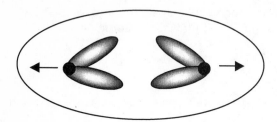

(a) Circle the correct stage for this cell:

    anaphase of mitosis,
    anaphase of meiosis I,
    anaphase of meiosis II,
    telophase of mitosis.

(b) How many picograms of chromosomal DNA would you expect in the cell shown above?_____

> **Concepts: chromosome mechanics (meiosis), symbolism, meiotic nondisjunction sex-linkage**

Chs. 8, 9, 10 Ques.22 The genes for *singed bristles (sn)* and *miniature wings (m)* are recessive and located on the X chromosome in *Drosophila melanogaster*. In a cross between a singed-bristled, miniature-winged female and a wild type male, all of the male offspring were singed-miniature.

> **(a)** Draw meiotic metaphase I chromosomal configurations which represent the X and/or Y chromosomes of the parental (singed-miniature female and wild type male) flies. Place gene symbols (*sn, m*) and their wild type alleles (*sn⁺, m⁺*) on appropriate chromosomes.

(b) Draw a mitotic metaphase chromosomal configuration that represents the X and/or Y chromosomes of the F₁ male. Place gene symbols (*sn, m*) on appropriate chromosomes.

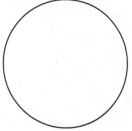

(c) Most of the female offspring from the above-mentioned cross were phenotypically wild type; however one exceptional female was recovered which had singed bristles and miniature wings. Given that meiotic nondisjunction accounted for this exceptional female, would you expect it to have occurred in the parental male of parental female?

(d) Draw a meiotic, labeled (with gene symbols) circumstance and division product(s) which could account for the exceptional female described above in part (c). (Confine your drawing to X chromosomes only).

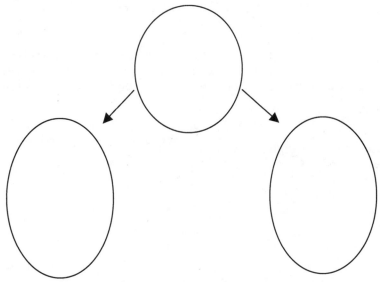

Sample Test Questions

> **Concepts: meiosis and chromosome numbers, DNA content**

Ch.8 Ques.23 The red fox (*Vulpes vulpes*) has 17 pairs of somewhat long chromosomes. The Arctic fox (*Alopex lagopus*) has 26 pairs of somewhat shorter chromosomes.

(a) If a female red fox is crossed with a male Arctic fox, what will be the chromosome number in the somatic tissues of the hybrid?

(b) Assume that a somatic G1 nucleus of the Arctic fox contains 12 picograms of DNA while a somatic G1 nucleus of the red fox contains 8 picograms of DNA. How much nuclear DNA could you expect in a G2 somatic nucleus of the hybrid?

> **Concepts: sex-linked inheritance (X-linked), pedigree construction, probability (product rule)**

Chs. 9, 11 Ques.24 Red-green colorblindness is inherited in man as an X-linked, recessive gene. Using the symbols below, draw a pedigree which is consistent with the following statements.

A phenotypically normal woman is married to a phenotypically normal man. The woman's parents are phenotypically normal but her maternal grandfather is colorblind. The woman's paternal grandparents as well as her maternal grandmother are phenotypically normal.

male = ☐
female = ◯
Rg = normal color sight
rg = colorblind

What is the probability that the first son born to the woman will be phenotypically normal (not be colorblind)?

Sample Test Questions

> Concepts: sex-linked inheritance (X-linked), chromosome mechanics, meiosis, conventional symbolism

Chs. 9, 11 Ques.25 In a *Drosophila* experiment a cross is made between a homozygous wild type female and a tan-bodied (mutant) male. All the resulting F_1 flies were phenotypically wild type. Adult flies of the F_2 generation (from a mating of the F_1's) had the following characteristics:

Sex	Phenotype	Number
Male	wild	346
Male	tan	329
Female	wild	702

(a) Using conventional symbolism, illustrate the genotype, *on an appropriate chromosomal configuration*, of a secondary oocyte nucleus of one of the F_1 females. Be certain to distinguish the X chromosomes from the autosomes. Note: *Drosophila melanogaster* has a diploid chromosome number of 8.

(b) Using the same conventional symbolism, give the genotype *on an appropriate chromosomal configuration* of a primary spermatocyte of the tan-bodied F_2 males.

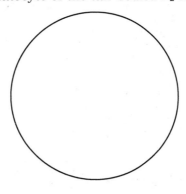

267

Sample Test Questions

> **Concepts: Mendelian genetics, monohybrid cross, dominance/recessiveness, 3:1 and 1:1 ratios**

Ch.9 Ques.26 *Gray* seed color in peas is dominant to *white*. Assume that Mendel conducted a series of experiments where plants were crossed and offspring classified according to the table below. What are the most probable genotypes of each parent?

Parents			Progeny	
			gray	white
(a) gray	X	white	81	79
(b) gray	X	gray	120	42
(c) white	X	white	0	50
(d) gray	X	white	74	0

(a) gray _____ X white_____
(b) gray _____ X gray _____
(c) white _____ X white _____
(d) gray _____ X white _____

> **Concepts: sex-linkage (X-linked), autosomal inheritance, dihybrid situation incomplete dominance, complete dominance**

Chs. 9, 10, 11 Ques.27 Hemophilia (type A) is recessive and X-linked in humans whereas the ABO blood groups locus is autosomal. Assume that the following matings were examined for the transmission of these genes. Give the expected phenotypes and numbers assuming 800 offspring are produced.

Group A: Females heterozygous for hemophilia with blood type AB mated to normal males with blood group O.

Group B: Females heterozygous for hemophilia with blood type AB mated to males with hemophilia and blood group AB.

268

Sample Test Questions

Concepts: Mendelian patterns, 1:1:1:1 ratio, null hypothesis, expected values
χ^2 analysis, interpretation of χ^2

Ch.9 Ques.28 For the cross *PpRr* X *pprr* where complete dominance and independent assortment hold, assume that you received the following results and you wished to determine whether they differ significantly (in a statistical sense) from expectation.

PR phenotypes   = 40
Pr phenotypes   = 10
pR phenotypes   = 20
pr phenotypes   = 30

(a) State the null hypothesis associated with this test of significance.

(b) How many degrees of freedom would be associated with this test of significance?

(c) Assuming that a Chi-Square value of 20.00 is arrived at in this test of significance, do you accept or reject the null hypothesis?

Degrees of Freedom	$p = 0.05$
1	3.84
2	5.99
3	7.82
4	9.49
5	11.07

Concepts: sex-linked inheritance (X-linked), gynandromorph production,
sex determination in Drosophila, insect development mitosis,
nondisjunction

Chs.8, 9, 10, 11 Ques.29 Explain the processes, genotypic, chromosomal, and developmental, which would lead to a bilateral gynandromorph in *Drosophila melanogaster* in which the male half of the fly has white eyes and singed bristles, while the female half is phenotypically wild type.

Sample Test Questions

> **Concepts: crossing over mechanisms, meiosis, gene mapping, chromosome mechanics**

Chs. 2,12 Ques.30 Assume that there are 18 map units between two loci in the mouse and that you are able to microscopically observe meiotic chromosomes in this organism. If you examined 150 primary oocytes, in how many would you expect to see a chiasma between the two loci mentioned above?

> **Concepts: linkage and crossing over, computation of map units, complete linkage independent assortment, lack of crossing over in males**

Ch.12 Ques.31 Given below are four dihybrid crosses between various strains of *Drosophila*. To the right of each are map distances known to exist between the genes involved. For each cross give the phenotypes of the offspring and the percentages expected for each.

	Matings	
Female	*Male*	*Map distance*
(a) *AB/ab*	*ab/ab*	20
(b) *Pq/pQ*	*pq/pq*	50
(c) *DB/db*	*db/db*	0
(d) *ab/ab*	*AB*/ab	20

(a) *AB/ab* X *ab/ab* _____

(b) *Pq/pQ* X *pq/pq* _____

(c) *DB/db* X *db/db* _____

(d) *ab/ab* X *AB*/ab _____

Sample Test Questions

> **Concepts: recombination in bacteria, transduction, transformation, lysogeny**

Ch.15 Ques.32 Below are phrases which refer to various forms of recombination in bacteria. For each, clearly state whether you **agree** or **disagree**. If you disagree, briefly explain your reason(s).

(a) Transduction is the process in which exogenous DNA is drawn into bacteria as a single-stranded structure, then integrated into the bacterial chromosome. [   ]

(b) Temperate phage are capable of entering a lysogenic cycle such that their genomes are incorporated into the bacterial chromosome. [   ]

(c) During the lysogenic cycle, phage are capable of producing bacteria when exposed to U.V. light. [   ]

> **Concepts: extranuclear inheritance, maternal effects**

Ch.14 Ques.33 Direction of shell coiling in the land snail, *Limnaea peregra*, is determined by alleles at a single locus: *dextral* (right) = *DD or Dd; dd = sinistral* (left). However, a maternal effect is present such that the genotype of the mother determines the direction of coiling (phenotype) of the immediate offspring. Given the following crosses, write the genotypes and phenotypes in the spaces provided. Be certain to indicate which genotypes go with which phenotypes.

	Source of sperm	**Source of egg**
Cross #1	**DD**	**dd**

Offspring genotype(s): [   ] *Offspring phenotype(s):* [   ]

	Source of sperm	**Source of egg**
Cross #2	**dd**	**Dd** (sinistral)

Offspring genotype(s): [   ] *Offspring phenotype(s):* [   ]

	Source of sperm	**Source of egg**
Cross #3	**Dd**	**Dd**

Offspring genotype(s): [   ]   *Offspring phenotype(s):* [   ]

271

Sample Test Questions

> **Concepts: genetic regulation, positive vs. negative control**

Ch.19 Ques.34 Depending on the regulatory system, in prokaryotes when the regulatory protein **is** or **isn't** bound to the DNA, the operon may be **on** or **off**. Fill in the chart below and give a brief explanation of your reasoning.

Relationship of Regulator Protein to DNA	*Operator*	
	Positive control	*Negative control*
is bound		
isn't bound		

> **Concepts: experimental strategies, action of nucleases, DNase restriction endonucleases**

Ch.16 Ques.35 DNase is often used to map the locations where DNA-binding proteins (histones, RNA polymerases, transcription factors, *etc.*) interact with DNA. Restriction endonucleases are used to cut DNA for identification and cloning. Why are these two different enzyme classes used in these different ways?

Sample Test Questions

Concepts: experimental strategies cDNA probes Southern blots electrophoresis
restriction endonuclease analysis hybridization

Ch. 16 Ques.36 Assume that you have a cDNA clone for the gene causing retinoblastoma, and you prepare Southern blots probing DNA in cells from normal individuals and from children with retinoblastoma. Genomic DNA is prepared using the restriction endonuclease *Hind*III (the *Rb* gene contains four *Hind*III fragments as indicated below), and the following hybridization appears:

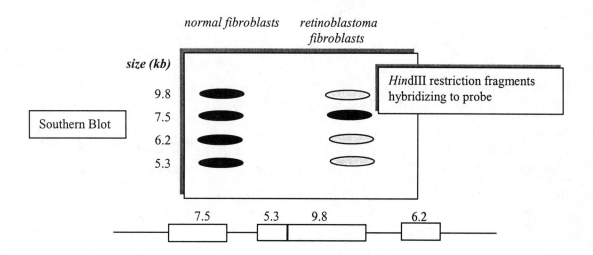

What does each band represent?

What conclusions can be drawn from these data?

Sample Test Questions

Concepts: experimental strategies electrophoresis restriction digestion fragment analysis

Chs. 16, 18 Ques. 37 The *thioredoxin* gene in bacteria aids in the necessary reduction of proteins. It encodes a protein of 108 amino acids and is contained in a 0.9 kb (*PstI/Bam*HI) fragment. The gene for kanamycin resistance is contained in a 1.4 kb (*Bam*HI/*Pst*I) fragment. The restriction map (one orientation) of these two genes (flanked by *Bam*HI sites) in a plasmid vector is presented below.

(A)

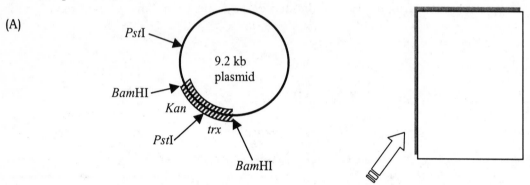

(a) Assume that the plasmid is restricted with the enzyme *Bam*HI. What would be the electrophoretic pattern of the cleaved fragments?

(b) Assume that the orientation given above is only a guess and that the *Bam* HI fragment containing the *trx* and *Kan* genes could possibly exist in the opposite orientation (B). What experiment would you perform to determine whether the orientation is as in (A) or (B)?

(B)

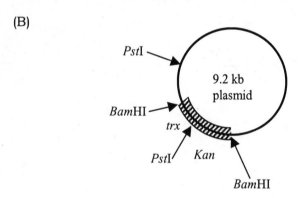

274

Sample Test Questions

Concepts: electrophoresis, DNA sequencing

Ch.16, 18 Ques.38 The Maxam and Gilbert DNA sequencing procedure involves chemical reactions (methylation) and cleavages (piperidine) that produce ^{32}P-labeled DNA fragments. The chemical reagents can give rise to G, G+A, C, and C+T cleavages. These fragments are then separated on electrophoretic gels, and the sequence is read directly from the gel. A sample gel is given below. From this gel, provide the sequence of the DNA fragment.

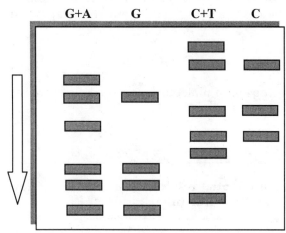

G+A   G   C+T   C

Small fragments at this end of gel

Concepts: determination, differentiation,, relationships, *Drosophila* development

Ch.21 Ques.39 Two terms, *determination* and *differentiation*, are consistently used in discussions of development.

(a) Provide a brief definition of each term.

(b) Which, determination or differentiation, comes first during development of *Drosophila*, for example?

Concepts: variable gene activity hypothesis, genomic equivalence evidence for differential transcription

Ch.21 Ques.40 Development may be defined as the attainment of a differentiated state. Given that all cells of a eukaryote probably contain the same complete set of genes, how do we currently explain development in terms of gene activity? What evidence supports your explanation?

Sample Test Questions

> **Concepts: immunoglobin structure, gene recombination, association of chains**

Ch.23 Ques.41 (a) Describe the general structure of an antibody molecule and the manner in which antibody variability is established.

(b) Assuming that a light chain has 300 V and 10 J regions and the heavy chain has 300 V, 10 D, and 10 J regions, about how many different immunoglobin light-chain genes and how many different heavy-chain genes could theoretically be formed in this organism? Assume that no nucleotides are lost or gained during recombination within the chains.

(c) How many different immunoglobin molecules could be generated from these heavy and light chains?

> **Concepts: interaction of gene products, relationship between genotype and phenotype, multi-factor inheritance**

Chs.24 Ques.42 Describe and exemplify similarities and differences between *discontinuous* and *continuous* traits at the *molecular* and *transmission* levels.

> **Concepts: Hardy-Weinberg applications to X-linked gene maintenance of gene frequencies over time**

Ch.25 Ques.43 Assume that in a particular population, approximately 8% of the males show red-green colorblindness. Knowing that this form of colorblindness is X-linked, what percentage of the females would be expected to be colorblind?

What would be the expected frequency of heterozygous females?

Assuming that the Hardy-Weinberg equilibrium assumptions pertain, what percentage of men will be colorblind in the next generation?

Sample Test Questions

> **Concepts: factors which change gene frequencies, influence of inbreeding on gene frequencies**

Ch.25, 26 Ques.44 List and briefly describe factors which change gene frequencies in populations. Is inbreeding a factor in changing gene frequencies? Explain.

> **Concept: relationship between speciation and Hardy-Weinberg assumptions**

Ch.26 Ques.45 A *species* is often defined as a population of interbreeding or potentially interbreeding organisms reproductively isolated from other such populations. Given such an isolated population, will speciation (formation of a new species) occur if the Hardy-Weinberg assumptions are met?

> **Concepts: natural selection, speciation**

Ch.26 Ques.46 Assume there exists a group of organisms which is temporally or spacially isolated from other such groups. This group has evolved genetic differences from neighboring groups. Describe process of speciation in terms of this genetically distinct group of organisms.

Sample Test Answers

Ch.1 Answer 1 lack of understanding of the genetic basis of variation and inheritance

Ch.2 Answer 2 Gregor Mendel

Ch.1 Answer 3 Genetics is a subdiscipline of biology concerned with the study of heredity and variation at molecular, cellular, developmental, organismal, and populational levels.

Ch.1 Answer 4 DNA or deoxyribonucleic acid is the hereditary material in eukaryotes and prokaryotes. Either DNA or RNA (ribonucleic acid) serves in viruses.

Ch.1 Answer 5 homologous

Ch.1 Answer 6 In mitosis, chromosome number remains constant, while in meiosis, chromosome number is reduced by half in the final products. In meiosis, there is pairing of homologous chromosomes.

Ch.1 Answer 7 nitrogenous bases, phosphate, deoxyribose sugar

Chs. 2, 3 Answer 8 (a) DNA replicates in a semiconservative manner such that each daughter strand is "half-new" and 'half-old" in a particular pattern.

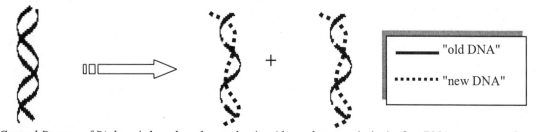

(b) The *Central Dogma of Biology* is based on the production (through transcription) of an RNA messenger from a DNA template and the subsequent "decoding" of that messenger by the process of translation.

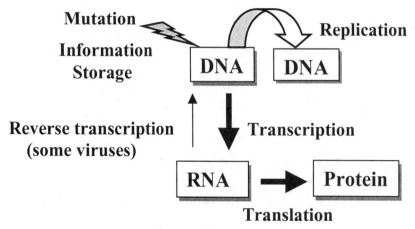

Sample Test Answers

(c) The DNA template contains a sequence of nitrogenous bases which specifies a code from which amino acids are ordered in proteins. Through tautomeric shifts and a number of other natural factors (radiation, chemicals), changes can occur in that sequence of bases. Indeed, the mechanisms by which genes replicate themselves generate errors and leaves us with the conclusion that DNA is an inherently unstable molecule.

(d) Given the variety of organisms and the variation within organisms, there must be numerous, hundreds of millions, elementary factors which are inherited. DNA can provide for this variety by differences in the length and sequence of bases for each inherited functional unit. Given that there are four different types of bases, a sequence having merely ten bases would be capable of 4^{10} (over 1 million) different sequences.

Common errors for Question 8: There are usually very few problems with this type of question, except that students often have difficulty clearly explaining that which they know in model form. Written descriptions tend to be more lists of examples rather than explanations of structures and/or processes.

Ch. 2, 3 Answer 9 As discussed in the text, the rate of reassociation of melted DNA increases as the proportion of repetitive DNA increases. Such relationships are reflected in $C_o t$ curves in which one plots the fraction of DNA reassociated against a logarithmic scale of $C_o t$ values which have the units (mole X sec/liter). Because denatured DNA strands which are repetitive have a higher likelihood of complementary interaction, the time of reassociation is less.

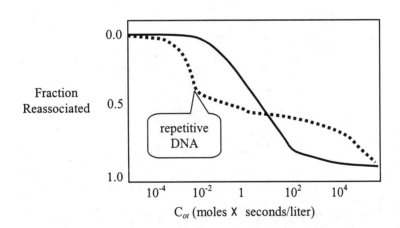

Common errors for Question 9: orientation (coordinates) of graph, significance of curve components, repetitive DNA fraction, unique fraction

279

Chs. 2, 3 Answer 10 (a) In this type of drawing, the various carbons of the sugar are readily apparent. It is the orientation and carbon numbering on the sugar which determines the 5'-3' orientation of the molecule. The bottom of the polymer has the 2' and 3' carbons projecting, while the top has the 5' carbon projecting. Therefore, the bottom, near the circle is the 3' end of the molecule.

(b) Notice that there is no vertical line protruding from the 2' carbon position in the drawing and that uracil (U) is present. The molecule must therefore be an RNA. **(c)** It is best to start this portion of the problem by roughly drawing the complementary strand, remembering that it will be antiparallel.

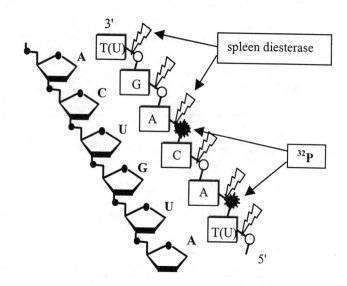

If it is a DNA complement, it will have thymine in place of uracil. There is no indication as to the complementary strand being RNA or DNA. Since all ATP's (or dATP's) have at their innermost phosphate a ^{32}P, make certain that they are properly labeled. Since spleen diesterase cleaves between the phosphate and the 5' carbon, the 5' neighbors (C, T or U) will be labeled with the ^{32}P.

(d) Since snake venom diesterase cleaves at the 3' position (between the phosphate and the 3' carbon), the originally labeled ATP (or dATP) will retain the label.

Common errors for Question 10: 5' to 3' orientations, labeling of complementary strand, understanding enzyme cleavages

Ch. 3 Answer 11 Even though circular, in the context of this question, DNA from *E. coli* can be viewed in the following manner. Replication will occur semiconservatively and give the following sedimentation profile.

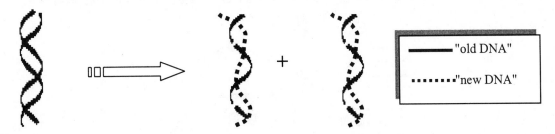

The heat treatment will cause the double-stranded structures to separate, giving the following strands and the profile as shown above.

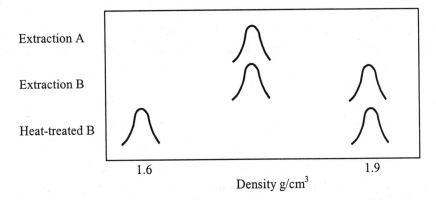

Common errors for Question 11: confusion with the labeling experiment, difficulty in seeing sedimentation profiles, application of semiconservative replication

Ch. 3 Answer 12 (a) There will be two telocentric metaphase chromosomes in the drawing and each chromatid will be labeled.

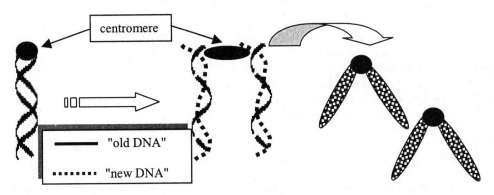

This autoradiographic pattern results because of semiconservative replication. Any cell which contains labeled chromosomes will have had its chromosomes pass through an S phase in the presence of label.

(b) Consider that the DNA was labeled with a dCTP having the innermost phosphate labeled. As this triphosphonucleoside is incorporated into the DNA, it will have the following relationship to its neighbors. *Snake venom diesterase* cleaves DNA at the 3' position meaning that it breaks the bond between the phosphate and the 3' carbon.

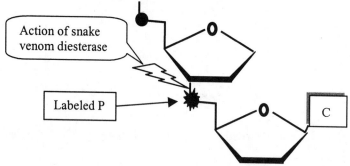

Therefore, the labeled phosphate remains attached to the 5' carbon of the cytosine nucleotide. The A+T/G+C ratio of 1.67 is of no consequence in answering this problem because all of the label remains attached to the cytosine.

Adenine_____   Guanine_____

Thymine_____   Cytosine <u>100%</u>

> **Common errors for Question 12:** inappropriate labeling pattern, inability to draw telocentric chromosomes, understanding cleavage at 3' position, eliminating extraneous information

Ch. 3 Answer 13

1. This must be the *5' end* of the polymer because all synthesis of polymers is 5' to 3' and the head of the arrow is at the other end of the polymer.
2. The *leading strand* is that strand which is synthesized continuously.
3. A *DNA ligase* will join the nicks.
4. The *lagging strand* is the discontinuous strand.
5. The free end that is complementary to the 5' end at arrow #1 must be the 3' end. That being so, the complement to that 3' end would be 5'. Therefore the *5' end* is at arrow #5.
6. A *helicase* is involved in unwinding the DNA helix.
7. An *RNA primer* is synthesized to initiate DNA synthesis.
8. *DNA gyrase* functions to remove supercoils generated by unwinding the DNA helix.
9. *Single-stranded binding proteins* stabilize the template which is to be replicated.

> **Common errors for Question 13:** determination of 5', 3' polarity, naming of enzymes involved

Sample Test Answers

Chs. 5, 6 Answer 14 Overall, this is a drawing of simultaneous transcription and translation in which the RNA polymerase is moving from top to bottom, making an mRNA which is complementary to one of the two strands of DNA. Ribosomes have added to the nascent (newly forming) mRNA and what appear to be polypeptide chains are protruding from the ribosomes.

(1, 2) In answering this question remember that all synthesis of nucleic acids starts at the 5' end and finishes at the 3' end. Therefore, immediately label the end near point "B" with a 5'. The projecting strand is the nascent mRNA. Recall that all orientation of complementary strands is antiparallel and since the RNA polymerase is going from top-to-bottom (according to the arrow) the end of the DNA strand from which the mRNA is copied is the 3' end. Now, since the 3' end of the DNA template strand is identified, its DNA complementary end (at point "A") must be the 5' end.

(3) The codon-anticodon relationship is also antiparallel (based on hydrogen bonding) and since the 5' end of the mRNA is identified, letter "C" must be at the 5' end.

(4) While the diagram is *not to scale*, given the folded structure of the molecule and its position in the ribosome, consider the RNA nearest to letter "D" as tRNA.

(5) Remember that the 3' end of the tRNA holds the amino acid, therefore letter "D" is where the amino acid would be attached.

(6) The tRNA binds mainly to the large subunit of the ribosome, while the mRNA binds mainly to the small subunit of the ribosome. The question asks for the S value of the rRNA in that small subunit. Simultaneous transcription and translation occurs in prokaryotes only (not eukaryotes). The S value for the small subunit of a prokaryotic ribosome is 30S but that value includes both rRNA and protein. The rRNA molecule however has an S value of 16, which is the correct answer.

(7) Since all of the ribosomes are moving along the same mRNA, the amino acid sequences are the same. Therefore, the amino acids nearest the letters "F" and "G" are the same.

> **Common errors for Question 14:** polarity of DNA and RNA strands, structure of ribosomes, overall understanding of translation

Chs.6, 7 Answer 15 One of the most frequent difficulties students have with this problem is remembering that the code is triplet and three bases in the mRNA code for each amino acid in a protein. The 5' end of the mRNA corresponds with the N-terminus of the amino acid chain.

(a) Transition mutations will cause amino acid substitutions. Counting over by "threes" from the N-terminus, the amino acid ile should be altered.

(b) Inserting a base between positions 3 and 4 will change the second amino acid; however, recall that acridine orange is a frameshift mutagen and insertion of a base will alter the reading frames for all "downstream" amino acids. Therefore all amino acids in positions two (glu) through twelve will be influenced (excluding degeneracy).

(c) Nitrous acid causes base substitutions, therefore a mutation in the 11th base would influence the amino acid leu.

> **Common errors for Question 15:** counting amino acids as bases, not understanding what base changes do to amino acid sequences

283

Sample Test Answers

Ch. 6, 7 Answer 16 Notice that there are two mutant strains (cannot grow on minimal medium) and one wild type strain (t410). The best way to approach these types of problems, especially when the data are organized in the form given, is to realize that the substance (supplement) which "repairs," as indicated by a (+), a strain is after the metabolic block for that strain. In addition, and most importantly, the substance which "repairs" the highest number of strains either is the end product or is closest to the end product.   Looking at the table, notice that supplement B "repairs" both the mutant strains. Therefore, it must be at the end of the pathway or at least after all the metabolic blocks (defined by each mutation). Supplement A "repairs" the next highest number of mutant strains (1) therefore it must be second from the end. The pathway therefore would be as follows:

t409   r3

Precursor----{----> A --{----> B

To determine the locations at which the strains block the pathway through mutation, apply a similar logic. A block that is "repaired" by all the supplements must be early in the pathway. A block which is "repaired" by only one supplement must be late in the pathway. A supplement which does not "repair" a strain is before that strain's metabolic block. Be certain not to introduce any "blocks" for the wild type strain.

> **Common errors for Question 16:** inability to construct a pathway, failure to see how additives "repair" mutant phenotypes, difficulty in assigning metabolic blocks in pathways

Ch. 5 Answer 17

1. Hydrophobic cluster formed by interaction of hydrophobic amino acids.

2. α helix formed from hydrogen bonds between components of the peptide linkage.

3. Covalent, disulfide bonds formed between cysteine residues

4. The polar amino acids will tend to orient to the outside of the protein where the charged R groups will interact with water.

> **Common errors for Question 17:** nature of hydrophobic clustering, understanding of α and β structures polar side chains and hydrophilic interactions

Chs. 5, 6 Answer 18 In some viruses, overlapping genes present a mechanism for providing two and sometimes more protein products from a single stretch of DNA. In eukaryotes, different sets of introns may be removed, thus providing for a variety of protein products from a single section of DNA. This process is often called *differential* or *alternative splicing.*

> **Common errors for Question 18:** Students often have difficulty in orienting *specific information* they have learned to a general question. If asked about overlapping genes, or differential hnRNA splicing, they would be able to develop an answer. Students sometimes confuse overlapping genes with the non-overlapping code.

284

Ch. 8 Answer 19 This question is intended to determine your understanding of mitosis, chromosome morphology, symbolism, the positioning of genes on chromosomes, and the changes in DNA content through the cell cycles. **(a)** Since the diploid chromosome number is six, there will be three bivalents, one involving metacentrics, and two involving acrocentrics in a primary oocyte. We can draw the chromosomes of the primary oocyte as follows:

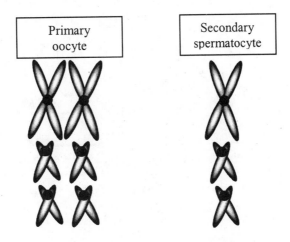

Since secondary spermatocytes arise after meiosis I, there should be only dyads, one metacentric and two acrocentric, and they should be aligned end-to-end as indicated in the above drawing .

(b) Given that there are about 20 picograms of DNA in a G1 nucleus, we would expect there to be 40pg in a G2 nucleus (after S phase) and 40pg to the point where homologous chromosomes separate in meiosis I. Secondary spermatocytes and secondary oocytes (as well as first polar bodies) should therefore, each have 20pg of DNA. After meiosis II the resulting nuclei should have 10pg each. If you understand events at interphase and in meiosis, this question is easy to answer. Carefully examine the figure below to understand events during the interphase as far as DNA content is concerned. Then examine the Figures in Chapter 2 in this book to see how chromosomes are behaving in meiosis. From this information you should see the answers as follows:

Primary spermatocyte	= 40 pg
First Polar Body	= 20 pg
Secondary oocyte	= 20 pg
Ootid (in G1)	= 10 pg

Sample Test Answers

It might be helpful to view changes in DNA content in graphic form:

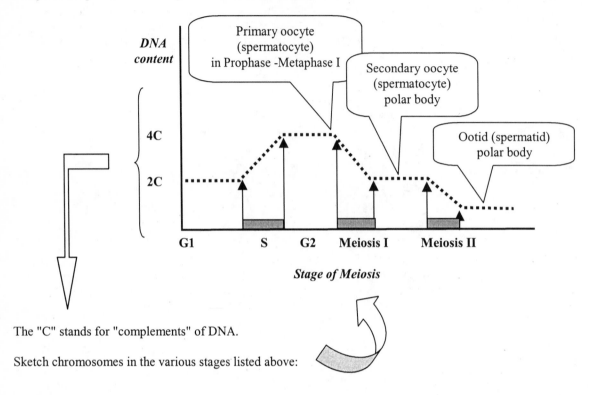

The "C" stands for "complements" of DNA.

Sketch chromosomes in the various stages listed above:

(c) If the mosquito is heterozygous for the recessive gene *wavy bristles* (*wb*) then it would have the genotype *Wb/wb*. Because there are four letters here representing the two genes, the slash between the symbols helps us to understand that there are only two genes being discussed.

We are asked to draw an acrocentric, mitotic metaphase chromosome complement in this heterozygous insect. We are expected to place the gene symbols on the chromosomes. Recall that there is no synapsis of homologous chromosomes in mitotic cells, therefore, the homologous chromosomes should not be placed side-by-side. Since sister chromatids are *identical* and homologous chromosomes are *similar*, we should draw the figure as follows:

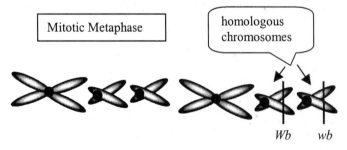

Common errors for Question 19: incorrect number of chromosomes, incorrect chromosome morphology, metacentric, acrocentric, poor relationship of DNA content to cells, inappropriate symbols, inappropriate placement of genes

286

Ch. 8 Answer 20 (a,b) Recall that a metacentric chromosome has "arms" of approximately equal length, while submetacentric and acrocentric chromosomes have arms of unequal length.

Metaphase I (Primary Oocyte): homologous chromosomes are replicated and synapsed. There will be two X chromosomes present because oocytes occur in females. On the metacentric chromosomes (#1), place, such that sister chromatids are identical, the *Dd*. Place the + - alternatives on the X chromosomes.

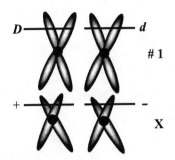

First Polar Body: the first polar body is a product of meiosis I, after homologous chromosomes have migrated to opposite poles. At this stage, dyads are present. Because females produce polar bodies, there should be an X chromosome present. Because the female is heterozygous, there are several possible answers. Note that there is only one representative of each allele for each gene pair.

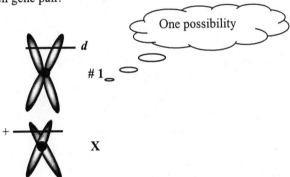

Secondary Spermatocyte: a secondary spermatocyte will have the same chromosome configuration as a First Polar Body. Both are products of meiosis I and dyads should be present. Because spermatocytes occur in males, there will either be an X chromosome or a Y chromosome present. There are, therefore, two possible answers. Regarding the genetic constitution of these cells, as stated in the problem, we are to assume that the male is Rh⁻ and + for the *G6PD* locus. Since this locus is on the X chromosome, only one genotype (regarding the X chromosome) can be presented.

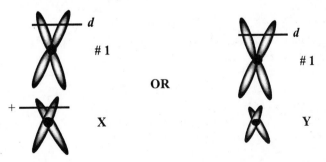

Sample Test Answers

Secondary Oocyte: being a product of meiosis I, dyads will be present. Because oocytes occur in females, there should be an X (not a Y) chromosome present. The genetic labeling pattern for the secondary oocyte will be the same as for the first polar body.

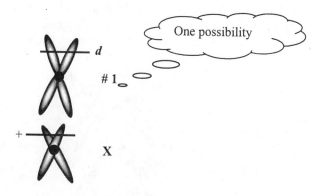

(c) If a G1 nucleus contains 6.5pg DNA, then the following DNA contents are expected.

Metaphase I (Primary Oocyte):	13pg
First Polar Body:	6.5pg
Secondary Spermatocyte:	6.5pg
Secondary Oocyte:	6.5pg

Common errors for Question 20: incorrect number of chromosomes, incorrect chromosome morphology, metacentric, acrocentric, poor relationship of DNA content to cells, inappropriate symbols, inappropriate placement of genes

---

Ch. 8 Answer 21 (a) Since the cell contains only two chromosomes ($2n=2$) and there are two chromosomes pictured, it can not represent a cell in the second phase (II) of meiosis. The chromosomes are telocentric which means that the centromere is at the end of the chromosome. When pulled at anaphase, two sideways "Vs" or "< >" would be expected and the cell would be at the anaphase stage of meiosis I. The only other possibility to produce the "< >" figure would be a metaphase chromosome at anaphase of mitosis or anaphase II of meiosis. However, these possibilities are negated because the chromosomes are stated as being telocentric.

(b) In order to get the correct answer for the second part one must consider that, because of the S-phase, at anaphase I the DNA complement is twice that of a G1 cell. Therefore the correct answer is 16pg DNA.

Common errors for Question 21: confusion on: significance of *telocentric* significance of chromosome number many students consider the chromosomes to be metacentric

---

Chs. 8, 9, 10 Answer 22 (a) The female parent would have the following labeled chromosomal symbolism remembering that at meiotic metaphase I, chromosomes are doubled, condensed, and synapsed.

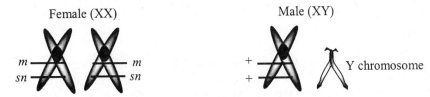

Female (XX) Male (XY)

(b) In general, mitotic metaphase chromosomes are not synapsed, although in *Drosophila* mitotic chromosomes do pair. To avoid confusion and to be consistent with what is expected in other organisms, the mitotic chromosomes will not be drawn in the paired state.

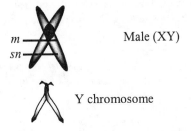

Male (XY)

Y chromosome

(c) All of the female offspring from the above cross should be heterozygous and phenotypically wild type. The one exceptional female could have resulted from maternal nondisjunction at meiosis I or II, thus producing an egg cell with two X chromosomes, each containing the *sn* and *m* genes. When fertilized by a sperm cell carrying the Y chromosome (along with the normal haploid set of autosomes) an $X^{sn\ m}X^{sn\ m}Y$ female is produced.

(d)

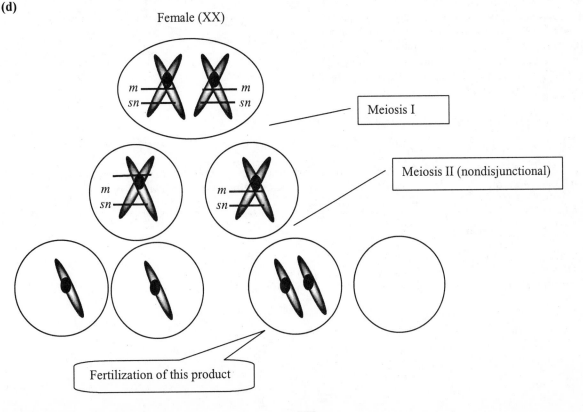

Female (XX)

Meiosis I

Meiosis II (nondisjunctional)

Fertilization of this product

Common errors for Question 22: incorrect chromosome morphology, X, Y chromosmes inappropriate symbols, inappropriate placement of genes, problems with meiotic nondisjunction

Ch. 8 Answer 23 (a) Since the chromosome numbers are given in *pairs*, recall that during meiosis each gamete contains one chromosome of each pair. If the red fox has 17 pairs of chromosomes, then each gamete will contain 17 chromosomes. For the Arctic fox, each gamete should contain 26 chromosomes. A zygote is produced from the union of the parental gametes, therefore it should contain 43 chromosomes (17 + 26). It turns out that some such hybrids are viable but usually sterile because of developmental and chromosomal alignment and segregational problems at meiosis.

(b) If G1 nuclei contain 12pg and 8pg DNA, then the gametes produced from these organisms will contain 6 and 4pg DNA respectively. Combining these gametes gives 10pg for a G1 cell. For a G2 cell there should be 20pg DNA.

Common errors for Question 23: confusion with pairs of chromosomes, gametic chromosome number, confusion with uneven number of chromosomes, confusion with *somatic* cells

Chs. 9, 11 Answer 24 Because the maternal grandfather was colorblind, the woman's mother is a carrier for this X-linked gene ($X^{Rg}X^{rg}$). The woman therefore has a 1/2 chance of inheriting the X^{rg} chromosome from her mother and a 1/2 chance of passing this X^{rg} chromosome to her son. The chance that the son will receive the X^{rg} chromosome is therefore 1/4 (1/2 X 1/2). However, the question asks for the probability that the son will be normal.

The answer is, therefore ,1 minus 1/4 which equals 3/4.

Common errors for Question 24: difficulty in setting up pedigree, inability to see independent probabilities, multiplication of independent probabilities, seeing that the *normal* is requested

Chs. 9, 11 Answer 25 (a) First one must determine whether the gene for *tan body* is X-linked or autosomal (not on the sex chromosome). Because half of the F$_2$ males are mutant and half are wild type, and all the females are wild, the gene for *tan body* is behaving as X-linked. The F$_1$ female is heterozygous, therefore she should have either of the alleles (t, t^+) on the one X chromosome (a secondary oocyte has one representative of each chromosomal pair) in the following arrangement. Because *Drosophila* has 8 chromosomes, each secondary oocyte should have four chromosomes (including the X chromosome).

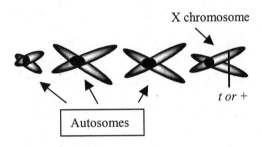

(b) A primary spermatocyte has the chromosomes in a doubled, condensed, and synapsed state. A tan-bodied male should have an X (containing a *t* gene) and Y chromosome as well as a diploid complement of autosomes.

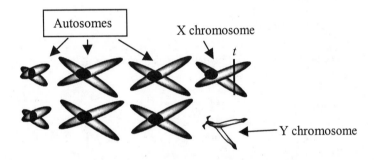

> **Common errors for Question 25:** difficulty in recognizing X-linked inheritance, problems with placing genes on chromosomes, problems with visualizing genome

Ch. 9 Answer 26 First, assign gene symbols: G = gray, gg = white

(a) Since there is an approximate 1:1 ratio in the progeny, the parental genotypes are *Gg* X *gg*.

(b) A 3:1 ratio is apparent, therefore the parental genotypes are *Gg* X *Gg*.

(c) Because there are no gray phenotypes and *gray* is the dominant allele, the parental genotypes must be *gg* X *gg*.

(d) Since there are no white types and the sample is sufficiently large, it is very likely that the parental genotypes are *GG* X *gg*.

> **Common errors for Question 26:** students usually have only minor problems with this type of question, some careless, random mistakes

Chs. 9, 10, 11 Answer 27 Set up the crosses with an appropriate symbol set such as the following:

h = hemophilia H = normal allele $I^A I^B$ = AB blood group $I^o I^o$ = O blood group.

Group A.

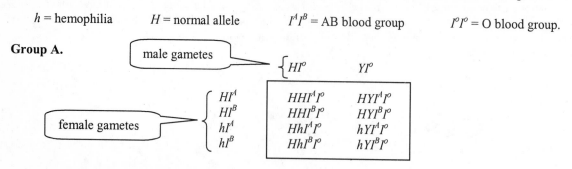

Collecting phenotypes gives:

1/4 female, normal, A blood (200)
1/4 female, normal, B blood (200)
1/8 male, normal, A blood (100)
1/8 male, normal, B blood (100)
1/8 male, hemophilia, A blood (100)
1/8 male, hemophilia, B blood (100)

Group B. In this example the forked-line method will be used. Consider what will be happening for the *hemophilia* locus independently from the blood group locus.

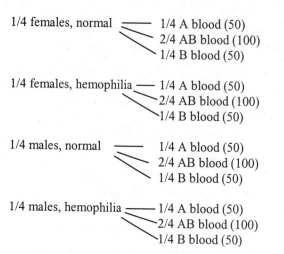

1/4 females, normal ——— 1/4 A blood (50)
                        2/4 AB blood (100)
                        1/4 B blood (50)

1/4 females, hemophilia — 1/4 A blood (50)
                        2/4 AB blood (100)
                        1/4 B blood (50)

1/4 males, normal ——— 1/4 A blood (50)
                      2/4 AB blood (100)
                      1/4 B blood (50)

1/4 males, hemophilia — 1/4 A blood (50)
                       2/4 AB blood (100)
                       1/4 B blood (50)

Common errors for Question 27: difficulty with any dihybrid situation, X-linked with autosomal inheritance, incomplete dominance, calculating frequencies, observed numbers

Ch. 9 Answer 28 (a) An appropriate null hypothesis for this example would be that the observed (measured) values do not differ significantly from the predicted ratio of a 1:1:1:1. One might also say that any deviation between the observed and predicted values is due to chance and chance alone.

(b) Because there are four classes being compared, there will be three degrees of freedom.

(c) Given that the Chi-square value of 20.00 is considerably greater than 7.82 (for three degrees of freedom) the null hypothesis should be rejected and the conclusion should be that the observed values differ significantly from the predicted values based on a 1:1:1:1 ratio.

> **Common errors for Question 28**: inability to see a 1:1:1:1 ratio, development of the expected ratios, interpreting probability values from table

Chs. 8, 9, 10, 11 Answer 29 In order for this type of fly to occur, the zygote must start out as a heterozygote in which both mutant genes are on one homologue and wild type alleles are on the other. In addition, one of the wild type chromosomes must get "lost" at the first mitotic division, thus making the female half $X^{+\ +}X^{w\ sn}$ and the other half $X^{w\ sn}$ O. The diagram below explains this situation.

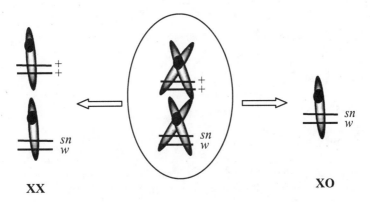

XX   XO

Because XX nuclei produce female tissue and XO nuclei produce male tissue, the phenotypes of the two sides are thus described. Developmentally, once the cleavage nuclei reach the peripheral areas of the egg to form a blastoderm, they become committed to their adult fate. Because there is little "wandering" of nuclei either during their migration to the egg periphery or after they reach the periphery, the male/female boundary is quite clean.

> **Common errors for Question 29 :** difficulty in setting up the problem, dealing with mitotic nondisjunction, embryonic development of *Drosophila*

Chs. 2, 12 Answer 30 The basis of the solution is to recall that crossing over occurs at the "four-strand stage" (after the S-phase) and each chiasma involves only two of the four chromatids present in each tetrad. Therefore for each chiasma only two of the four, or 1/2, of the chromatids are crossover chromatids.

Gene mapping basically is the process of dividing the number of crossover chromatids by the total number of chromatids. Since each chiasma involves only two of the four chromatids, the map distance must be half of the chiasma frequency. If there are 18 map units between two genes then the chiasma frequency would be 36%. If one examined 150 primary oocytes one would therefore expect to see 0.36 X 150, or 54 cells with a chiasma between the two loci.

> **Common errors for Question 30**: problem seeing relationships of chiasma and map units, problems "seeing" meiosis, visualization of crossing over

Ch. 12 Answer 31 The key to solving these types of "reverse mapping" problems is to keep in mind that a map unit is computed by the equation (# crossover types/total number (X 100)) and that if the map distance is given it is easy to determine the percentages of parental and crossover offspring. Remember that there are two classes of crossovers and two classes of parentals from each cross.

(a)   *AB/ab* = 40%, *ab/ab* = 40% (parentals)
             Ab/ab = 10%, *aB/ab* = 10% (crossovers)

(b)   *Pq/pq* = 25%, *pQ/pq* = 25% (parentals)
             PQ/pq = 25%, *pq/pq* = 25% (crossovers)

             Notice that this is independent assortment.

(c)   *DB/db* = 50%, *db/db* = 50% (all parentals)

(d)   *AB/ab* = 50%, *ab/ab* = 50% (all parentals)

The reason that there are all parentals and no crossovers in this cross is that there is no crossing over in male *Drosophila*. With no crossing over, the *AB/ab* chromosomes in the male are passed to gametes without crossovers.

> **Common errors for Question 31**: difficulty going from map units to offspring frequencies, failure to see that there are two parental and two crossover classes, careless mistakes

Ch. 15 Answer 32 (a) Disagree: The process being described refers to *transformation* not transduction. Transduction is *phage-mediated* recombination, whereas in transformation exogenous DNA is taken up as indicated in the statement.   **(b) Agree:** Viruses which can enter either the lytic or lysogenic cycle are called *temperate* viruses. During the process of lysogeny, the viral chromosome is integrated into the bacterial chromosome as stated. **(c) Disagree:** This statement is fairly silly in that it states that phage are capable of producing bacteria. Regardless of the exposure to U.V. light, phage can not produce bacteria. Ultraviolet light can cause induction of the lytic cycle, therefore phage, when lysogenic bacteria are exposed.

> **Common errors for Question 32**: carelessness in reading statements, confusion as to what terms mean: transduction, transformation, conjugation, lysogeny, temperate viruses

Sample Test Answers

Ch.14 Answer 33 The definition provided initially in the problem can be applied directly to the solution of this problem. The *genotype* of the mother determines the direction of coiling (phenotype) of the immediate offspring. In this case, one merely assigns the genotypes on the basis of normal Mendelian principles, then the phenotypes based on the genotype of the mother.

Notice in Cross 2, the *Dd* has a sinistral phenotype. While this may confuse some students, remember that the phenotype is determined by the *maternal* genotype. When early developmental events are involved, often the maternal genotype will have a significant influence over those events because the mother makes the egg.

Cross #1:	*Offspring genotype(s):*	*Offspring phenotype(s):*
	Dd	all sinistral

Cross #2:	*Offspring genotype(s):*	*Offspring phenotype(s):*
	Dd	all dextral
	dd	

Cross #3:	*Offspring genotype(s):*	*Offspring phenotype(s):*
	DD	all dextral
	Dd	
	dd	

> **Common errors for Question 33:** When students are reminded of the nature of a maternal effect, that is, given a definition, there are very few errors. However, when they are asked to work the problem without the definition given, many have difficulty. Cross #2 causes most of the problems.

Chs.19 Answer 34 Any time a regulatory protein interacts with DNA and transcription is stimulated, it is called *positive* control. Any time a regulatory protein interacts with DNA and represses transcription it is called *negative* control. In completing the chart, apply these simple rules.

Relationship of Regulator Protein to DNA	*Operator*	
	Positive control	*Negative control*
is bound	**on**	**off**
isn't bound	**off**	**on**

> **Common errors for Question 34:** confusion with positive and negative control confusion with repressible and inducible systems

295

Sample Test Answers

Ch. 16 Answer 35 DNase is a general term which includes a variety of exo- and endonucleases which cleave DNA from the ends or internally, respectively. Such cleavage is often irrespective of base sequence. If naked DNA is exposed to DNases, it is rapidly degraded to oligo- and mononucleotides. When protein is associated with DNA, it protects regions from degradation. Such protected regions can be analyzed as to base content. Restriction endonucleases cleave DNA at specific sequences often hundreds or thousands of base pairs apart. If one is interested in mapping protein binding sites, one would want to use an enzyme (a DNase) with frequent yet relatively random cleavage characteristics, not restriction endonucleases.

Common errors for Question 35: understanding overall strategy, differences between DNases, restriction endonucleases

Ch.16 Answer 36 Notice that the fragment sizes (in kb) on the left side of the figure match the *Hind*III restriction fragments which are hybridizing (cDNA probe + genomic fragment) to the radioactive probe. Each band therefore represents a region where the radioactive probe is "trapped" by complementary base pairing to single-stranded DNA fragments which are bound to the filter. The smaller fragments migrate faster in the gel and therefore are in the bottom portion while the larger fragments are at the top, near the origin. Notice that the intensity of the bands from the normal individual is somewhat uniform, indicating that all the restriction fragments are found in equal amounts. However, the intensity of three of the bands from the patient with retinoblastoma are about half as dense as in the normal. One band (7.5 kb) has the same intensity as in the normal. Because humans are diploid organisms, with normally two copies of each gene, one may hypothesize that the individual with retinoblastoma has a heterozygous deletion of a portion of the retinoblastoma gene which includes *Hind*III fragments (9.8, 6.2, and 5.3 kb). It is likely that the inheritance of such a deletion is instrumental in causing familial retinoblastoma.

Common errors for Question 36: understanding of experimental design, cDNA probes, Southern blots electrophoresis, restriction endonuclease analysis, hybridization, recognition of deletion

Chs. 16, 18 Answer 37 Below is a drawing of the expected product if either of the above plasmids (A) or (B) is restricted to completion with *Bam*HI. There should be a 2.3 kb fragment (1.4 + 0.9 kb) and the remainder (9.2 - 2.3 = 6.9). Notice that the 6.9 kb fragment migrates slower (higher in the gel) than the 2.3 kb fragment. Also notice that the intensity of the stain is less in the smaller band because there is less DNA to bind the stain.

(a)

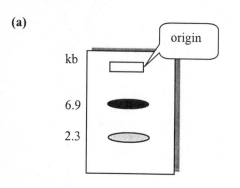

(b) To distinguish between the (A) and (B) orientations, one could make use of the change in position of the *Pst*I restriction site in the two orientations. First, estimate the number of kb in the two fragments resulting from *Pst*I restriction of orientation (A). Notice that in orientation (A) the *Pst*I fragments are approximately 2.4 (1.4 for the *Kan* gene + about 1.0) kb and 6.8 kb. In orientation (B) the sizes would be approximately 1.9 (0.9 for the *trx* gene + about 1.0) and 7.3 kb. With appropriate standards, these size differences could be distinguished on agarose gels.

Common errors for Question 37: electrophoretic analysis, restriction enzyme analysis, experimental design

Chs.16, 18 Answer 38 It is a relatively simple procedure to determine the sequence of DNA from the gel given. Start at the bottom with the smallest fragments. As one reads up the gel, one is reading from the 5' to the 3' direction. In the first case, note that there is a band in both the G+A and G lanes. Read this as a "G" because if an "A" occurs, it would exist as a band *only* in the G+A lane. The sequence would be as follows: 5'-GTGGTCACGACT

> **Common errors for Question 38:** reading the gel as the sequence, difficulty in dealing with G+A and C+T lanes, careless mistakes

Ch. 21 Answer 39 *Determination* is a significant, complex, yet poorly understood process whereby the specific pattern of genetic activity is initially established in a cell. This pattern will direct the developmental fate (differentiation) of that cell.

Differentiation is the process of cellular expression of the determined state. It is the complex series of genetic, morphological, and physiological changes which characterize the variety of adult cells.

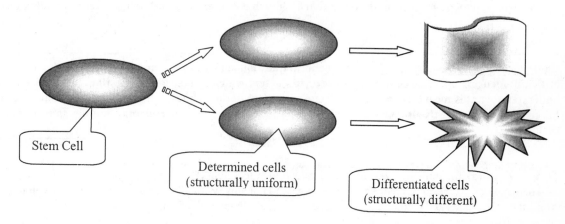

Stem Cell / Determined cells (structurally uniform) / Differentiated cells (structurally different)

(b) *Determination* occurs before *differentiation*. In *Drosophila*, determinative events are thought to occur about the time of blastoderm formation, when nuclei encounter the peripheral regions of the egg. *Differentiation* of most of the adult cells occurs during metamorphosis, some five to six days after embryogenesis (determination).

Blastoderm (embryogenesis)   *Pupation (morphogenesis)*

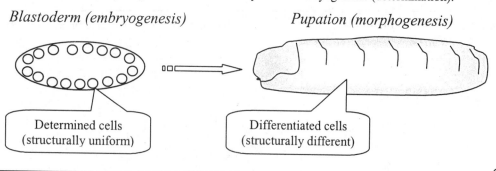

Determined cells (structurally uniform) / Differentiated cells (structurally different)

> **Common errors for Question 39:** providing accurate definitions, failure to relate determination and differentiation to each other temporally

Sample Test Answers

Ch. 21 Answer 40 The *variable gene activity hypothesis* of differentiation acknowledges the genomic equivalence of cells within an organism and assumes that of all the genes in a given cell type, only certain ones produce products while the others are shut down and are not transcribed. Certain genes will be active in all cells, those *housekeeping* genes coding for vital cellular functions, while others will be differentially regulated in various cell types. Differential gene transcription occurs in both spatial (different cells of an organism) and temporal (different times during development) dimensions.

Support for this model is provided by several observations and experiments.

1. *Chromosome puffs*: Specific puff patterns, representing differential gene activity, are observed in dipteran polytene chromosomes at different times during development.

2. *Isozymes*: Differential gene activity is demonstrated by the observation that different forms of the same enzyme (isozymes) are present in cells of different tissues. This evidence assumes that such isozyme patterns are not caused by post-transcriptional forms of genetic regulation.

3. *Growth hormone*: *In situ* hybridization and immunochemical studies in mouse embryos demonstrates spatial and temporal aspects of the regulation of growth hormone transcripts in the anterior pituitary gland.

Common errors for Question 40: With a general question such as this, students sometimes have difficulty focusing on the area in their notes or in the text which relates to the question. Students may understand what is meant by the variable gene activity hypothesis but not immediately see that it relates to the question. Students also have difficulty in relating a variety of experimental findings to a general theme.

Ch. 23 Answer 41 (a) An antibody molecule contains two identical light (L) chains and two identical heavy (H) chains. There are two classes of L chains and five classes of H chains. Variability results from variability in the genes encoding the chains and recombination among genes making up each chain.

(b) L : 300 V X 10 J = 3,000, H: 300 V X 10 D X 10 J = 30,000.

(c) Total molecular variation: 3,000 L X 30,000 H = 9 X 10^7

Common errors for Question 41: source of variability, general antibody structure, randomness of recombination, careless computation

Sample Test Answers

Ch. 24 Answer 42 At the *molecular level* one would consider that in discontinuous inheritance the gene products are acting fairly independently of each other, thereby providing a 9:3:3:1 ratio in a dihybrid cross for example. Genotypic classes

$$A\_B\_, A\_bb, aaB\_, \text{ and } aabb$$

can be clearly distinguished from each other because the gene products from the *A* locus produce distinct influences on the phenotypes as compared to those gene products from the *B* locus. Exceptions exist where epistasis and other forms of gene interaction occur. In discontinuous inheritance one would consider each locus as providing a *qualitatively* different impact on the phenotype. For instance, even though the *brown* and *scarlet* loci interact in the production of eye pigments in *Drosophila*, each locus is providing qualitatively different input.

In continuous inheritance, we would consider each involved locus as having a quantitative input on the production of a single characteristic of the phenotype. In addition, although it may not always be the case, we would consider each gene product as being qualitatively similar. Under this model, the *quantity* of a particular set of gene products, influenced by a number of gene loci, determines the phenotypic characteristic.

At the *transmission level* one sees "step-wise" distributions in discontinuous inheritance but "smoother" or more bell-shaped distributions in continuous inheritance. For instance, in a dihybrid situation (*AaBb* X *AaBb*) where independent assortment holds, one would obtain a 9:3:3:1 ratio (assuming no epistasis, *etc.*) under a discontinuous mode but a 1:4:6:4:1 where genes (or gene products) are acting additively (continuous inheritance). Both patterns are formed from normal Mendelian principles of segregation, independent assortment, and random union of gametes. It is the manner in which the genes (or gene products) interact which distinguishes discontinuous from continuous inheritance.

> **Common errors for Question 42:** difficulty with "molecular level" of the question, confusion over differences and similarities relating discontinuous and continuous patterns

Ch. 25 Answer 43 Since 8% of the males express the trait and males have only one X chromosome, the frequency (*q*) of the recessive gene would be .08 and *p* would be .92. Since females have two X chromosomes, the expected frequency of females that are homozygous for the color blindness gene would be q^2 or .0064 (.64%).

The frequency of females that are heterozygous would be 2*pq* or 2(.08)(.92) = .1472 or 14.72%.

Because the population is in equilibrium, the frequency of men with color blindness will not change from generation to generation. Eight percent of the men will be color blind in the next generation. Students should be aware of many deviations of these types of questions. The basic scheme is the Hardy-Weinberg equilibrium and the equations which apply.

> **Common errors for Question 43:** application of the Hardy-Weinberg equations to an X-linked gene, failure to apply the Hardy-Weinberg equations in determining the frequency of heterozygous females, students often make the question harder than it is by forgetting that under equilibrium conditions, gene frequencies do not change

Ch. 25, 26 Answer 44 *Mutation*, while being an original source of genetic variability, is not usually considered to be a significant factor in changing gene frequencies.

Migration occurs when individuals move from one population to another. The influence of migration on changing gene frequencies is proportional to the differences in gene frequency between the donor and recipient populations. Organisms often migrate as a result of some stress. Those organisms suffering from the most stress are often those that leave. Therefore, they do not represent a random sample of the individuals in that home range.

Selection can be a significant force in changing gene frequencies. It results when some genotypic classes are less likely to produce offspring than others. Selection may be directional, stabilizing, or disruptive.

Genetic drift can be a significant force in changing gene frequencies in populations that are numerically small or have a small number of effective breeders. In such populations, random and relatively large fluctuations in gene frequency occur by "sampling error."

Inbreeding is not a significant factor in changing gene frequencies in populations, however, it will change zygotic or genotypic frequencies. The number of homozygotes will increase at the expense of the heterozygotes.

> **Common errors for Question 44:** failure to provide a complete list, failure to briefly and adequately describe each term, failure to see that inbreeding does not, in itself, change gene frequencies

Ch. 26 Answer 45 Reproductive isolation can occur because of the introduction of geographic barriers or other dramatic changes in the environment which subdivide a population. Such factors facilitate speciation because gene flow is eliminated or at least restricted. With gene flow restricted, isolated populations can experience changes in gene frequencies when the Hardy-Weinberg assumptions are not met. Under Hardy-Weinberg equilibrium conditions where there is *random mating, no genetic drift, no selection, no mutation*, and *no migration*, gene frequencies will remain the same and speciation will not occur.

> **Common errors for Question 45:** confusion as to what the question is asking, difficulty in relating information from one chapter to information contained in a different chapter

Ch. 26 Answer 46 Any circumstance or process which favors changes in gene frequencies has the potential of generating a new species.

Factors such as selection, migration, genetic drift, or even mutation, may be important in species formation. One would certainly include geographic and/or temporal isolation as major barriers to gene flow and thus an important process in such formation.

Natural selection occurs when there is non-random elimination of individuals from a population. Since such selection is a strong force in changing gene frequencies, it should also be considered as a significant factor in species formation.

> **Common errors for Question 46:** explanations of terms, general relationships among diverse phenomena